Guide for Design, Installation, and Assessment of Post-installed Reinforcements

Ray K. L. Su

Daniel T. W. Looi

Yanlong Zhang

HKU
PRESS
香港大學出版社

Hong Kong University Press
The University of Hong Kong
Pokfulam Road
Hong Kong
https://hkupress.hku.hk

ISBN 978-988-8528-60-8 (*Paperback*)

British Library Cataloguing-in-Publication Data
A catalogue record for this book is available from the British Library.

10 9 8 7 6 5 4 3 2 1

Printed and bound by Hang Tai Printing Co., Ltd. in Hong Kong, China

Disclaimer

The authors assume no responsibility for any injury, damage, liability, negligence, and/or otherwise to any individual or property from the use or application of any of the methods, products, instructions, or ideas contained in the material herein.

Contents

Figures

Tables

Preface

Post-installed reinforcements use adhesive or cementitious grout to bond the reinforcements and concrete together. They are widely used to connect new structural components to old concrete structures. However, design guidelines for post-installed reinforcements that specifically emphasise material quality and quality control are not available in Hong Kong.

Guide for Design, Installation, and Assessment of Post-installed Reinforcements (hereinafter the Guide) is the first of its kind in Hong Kong and provides a comprehensive summary of the most recent design theories, available adhesion systems, appropriate installation methods, and quality control of post-installed reinforcements. The highlights of the Guide are as follows:

(1) The Code of Practice for Structural Use of Concrete 2013 is applied to the most recent developments in post-installed reinforcement requirements under design frameworks in Europe.

(2) Six proposals are offered to circumvent the issues of long anchorage length, without compromising the structural safety of the connection.

(3) Practical design examples are provided for both simply supported and moment connection cases, including an advanced strut-and-tie method, which has been validated recently through testing carried out at the University of Hong Kong.

The Guide provides good practices for the design, installation, and quality control of post-installed reinforcements. It is a useful reference for designers, contractors, and building control bodies.

Ray K. L. Su
29 February 2020

Acknowledgements

The authors would like to express their appreciation for the support and advice from Dr Giovacchino Genesio and Mr Ken T. K. Au Yeung. We would also like to thank Ir Augustus Yuen Fai Lee, who contributed to the drafting of Example 4 in Chapter 4 and its relevant technical components in the Guide.

Abbreviations

ACI	American Concrete Institute
BA	Anchorage design for bonded anchors
BSI	The British Standards Institutions
CEN	European Committee for Standardization
EAD	European Assessment Document (documentation of the methods and criteria accepted in EOTA as being applicable for the assessment of the performance of a construction product in relation to its essential characteristics)
EOTA	European Organisation for Technical Assessment
ETA	European Technical Assessment (a document providing information about the performance of a construction product, to be declared in relation to its essential characteristics)
ETAGs	European Technical Assessment Guidelines (valid to 30 June 2013)
HKBD	Hong Kong Buildings Department
ICC-ES	International Code Council Evaluation Service, Inc.
MPII	Manufacturer's published installation instructions
RC	Reinforced concrete
RA	Anchorage design for reinforcing bar
STM	Strut-and-tie model
ULS	Ultimate limit state

Symbols

A_c	Cross-sectional area of concrete
A_i	Area of joint
A_s	Area of reinforcement that crosses interface, including ordinary shear reinforcement in shear friction calculations; cross-sectional area of reinforcement in other applications
$A_{s,rqd}$	Cross-sectional area of reinforcement required
$A_{s,rqd,m}$	Cross-sectional area of reinforcement required at mid-span
$A_{s,min}$	Minimum cross-sectional area of reinforcement
$A_{s,prov}$	Cross-sectional area of reinforcement provided
$A_{s,prov,m}$	Cross-sectional area of reinforcement provided at mid-span
$A_{s,simplified\ rules}$	Cross-sectional area of reinforcement due to assumed partial fixity in simplified rules of detailing
$A_{s,surface}$	Lateral surface area of steel bar bonded with concrete base material
A_{st}	Cross-sectional area of transverse reinforcement
$A_{st,min}$	Minimum cross-sectional area of transverse reinforcement
F_{bond}	Resistance of anchorage bond
F_{Ed}	Axial force calculated with strut-and-tie model
F_o	Compressive force of strut in strut-and-tie model
F_R	Compressive force capacity of strut in strut-and-tie model
F_{rebar}	Compression or tension force in steel bar
K	Factor for beams and slabs in EN 1992-1-1 (2004) in calculation for α_3
M_{Ed}	Bending moment
M_{sp}	Splitting moment
N_{Ed}	Axial force (direct axial or result of bending)
$\overline{N_u}$	Measured mean ultimate load
R_r	Relative rib area of tested reinforcing bar
S_d	Ultimate design load

V_{Ed}	Transverse shear force in shear friction calculations; design shear force in other applications
W_{sp}	Section modulus
a	Shear span
b	Width of concrete section
b_i	Width of interface
c	Factor that depends on roughness of interface in shear friction calculations; cover thickness in other applications
c_1	Thickness of side cover
c_d	Minimum concrete cover
c_s	Concrete cover
d	Effective depth of beam or slab
d_b	Nominal diameter of tested reinforcing bar
f_{bd}	Design bond stress under static loading used in EN 1992-1-1 (2004)
$f_{bd,p}$	Maximum bond strength in case of pull-out taken from relevant anchor approval
$f_{bd,seis}$	Design bond stress under seismic loading
f_{bu}	Factored bond stress capacity used in HKBD 2013
f_{cd}	Design cylinder strength of concrete
f_{ctd}	Design tensile strength of concrete, which is equal to characteristic tensile strength at 5% fractile with consideration of a partial safety factor ($f_{ctk,0.05}/\gamma_m = 1.5$)
$f_{ctk,0.05}$	Characteristic tensile strength of concrete at 5% fractile
f_{cu}	Characteristic cube strength of concrete
f_n	Stress per unit area caused by minimum external normal force across interface that can act simultaneously with shear force, positive for compression and negative for tension
f_o	Compressive stress of strut in strut-and-tie model
f_{rebar}	Stress experienced by reinforcement
f_{sd}	Design stress
f_{sp}	Splitting stress
f_{yd}	Design yield stress capacity of steel reinforcement connector
f_{yk}	Characteristic yield strength of steel reinforcement
f_{uk}	Nominal tensile strength of steel reinforcement
g_k	Dead load
h	Depth of section
h_{beam}	Depth of beam
h_{col}	Depth of column

h_{slab}	Depth of slab
h_{wall}	Thickness of wall
k_b	Reduction factor in Section 2.2.2 of EAD 330087-00-0601 (2018)
l_b	Embedment or anchorage length
l_{b1}	Effective anchorage length
l_{bd}	Design anchorage length
$l_{bd,sp}$	Design anchorage length due to splitting bond strength
$l_{b,min}$	Minimum anchorage length
l_{bn}	Design or effective anchorage depth
$l_{b,rqd}$	Required embedment length
$l_{beam,sr}$	Anchorage length in accordance to simplified rules for simply supported beam
l_{inst}	Installed anchorage depth
$l_{inst,max}$	Maximum installed hole length
l_n	Span length
$l_{slab,sr}$	Anchorage length in accordance with simplified rules for simply supported slab
l_{span}	Span length of slab
l_o	Design lap length
$l_{o,min}$	Minimum lap length
p	Transverse pressure
p_1	Percentage of lapped bar
q_k	Live load
s	Clear spacing of bars
u	Reinforcement perimeter
ν	Shear strength reduction factor for cracked concrete
v_{Ed}	Applied shear stress
v_c	Concrete shear stress capacity
$v_{c.enhanced}$	Enhanced concrete shear stress capacity near support
w_{strut}	Strut band width in strut-and-tie model
z	Lever arm of section
z_{lr}	Effective inner lever arm in strut-and-tie model
α	Inclination angle formed by longitudinal axis of reinforcement with contact interface, and restricted between 45° to 90°
α_1	Coefficient for effect of form of bars, assuming adequate cover
α_2	Coefficient for effect of minimum concrete cover that takes splitting failure into consideration

α_2'	Coefficient to consider for effect of minimum concrete cover
α_3	Coefficient to account for effects of confinement with transverse reinforcement
α_5	Coefficient to account for effect of pressure transverse to plane of splitting along design anchorage length
α_6	Coefficient of percentage of lapped bar (ρ_1) relative to total cross-sectional area within $0.65\ l_o$ from centre of lap length
α_{ct}	Coefficient to account for effect of long term and loading effects
α_{lb}	Amplification factor for minimum anchorage length
β	Ratio of longitudinal force in new concrete area and total longitudinal force in either compression or tension zones, both calculated for section considered in calculations for shear friction; bond stress coefficient according to HKBD 2013
β_{strut}	Strut efficiency factor in strut-and-tie model
δ	Factor for increase in design bond strength with increasing related concrete cover due to different types of mortar
γ_c	Partial safety factor for concrete according to EN 1992-1-1 (2004)
γ_m	Partial safety factor for material according to HKBD 2013
γ_s	Partial safety factor for steel according to EN 1992-1-1 (2004)
κ_1	Interaction coefficient for tensile force activated in dowel bar
κ_2	Interaction coefficient for flexural resistance
λ	Ratio of excess transverse reinforcement area to longitudinal reinforcement area
μ	Factor that depends on roughness of interface
η_1	Coefficient for bond conditions
η_2	Coefficient for influence of reinforcement diameter
ϕ	Diameter of reinforcement
ρ	Ratio between area of reinforcement (A_s) that crosses interface, including ordinary shear reinforcement and area of joint (A_i)
τ_{Edi}	Demand of shear stress at interface of concrete
τ_{Rdi}	Capacity of shear stress at interface of concrete
$\bar{\tau}$	Calculated mean tension bond strength
θ	Strut angle

1
Introduction of Post-installed Reinforcements

1.1   Scope of guidelines

Conventionally, reinforcements are placed in formwork prior to concrete pouring. These 'pre-installed' reinforcements used in conventional construction methods are known as cast-in reinforcements, which are used to form monolithic construction. However, in many current construction circumstances, reinforcements are post-installed in existing structures with alteration and addition (A&A) work or retrofitting work, or deliberately designed for the convenience of the construction sequence. Post-installed reinforcements are drilled and installed into cured concrete, which are bonded by using a qualified adhesion system on one side of the interface, and usually serve as starter-bars and/or used to create lap splices with the reinforcements in new concrete structures on the other side of the interface. Hence, post-installed reinforcements should not be confused with post-installed anchor bolt systems, as the latter are commonly used to connect concrete with steel structural or non-structural elements (i.e., concrete pedestals and steel column base plates or reinforced concrete (RC) slabs and handrails). The Guide provides the installation, design, and assessment guidelines for post-installed reinforcements. Readers are advised to refer to a local Hong Kong reference for post-installed anchor bolt systems (Cho and Chan, n.d.).

Some application examples of post-installed reinforcements are shown in Figure 1.1. New RC slabs or beams can be attached horizontally with post-installed reinforcements onto existing RC shear walls and columns (see Figure 1.1(a) on p. 2) or slabs (see Figure 1.1(b)). Figure 1.1(c) shows an example of an RC column that is cast vertically into a foundation with or without lap splices. New concrete can be overlaid for wall strengthening, column jacketing, and slab thickening by using post-installed reinforcement technology (see Figure 1.1(d)).

1.1.1 International guidelines for post-installed reinforcements

Despite the popular use of holistic design principles and practices in construction, those for post-installed reinforcement systems are not explicitly provided in modern international structural design codes (for example, EN 1992-1-1 (2004) and ACI 318 (2014)). However, the rationality of some of the design philosophies can be traced in codes based on the associated failure modes; for instance, the provisions for anchorage length design (i.e., Cl. 8.4 in EN 1992-1-1 (2004) and Chapter 25 of ACI 318

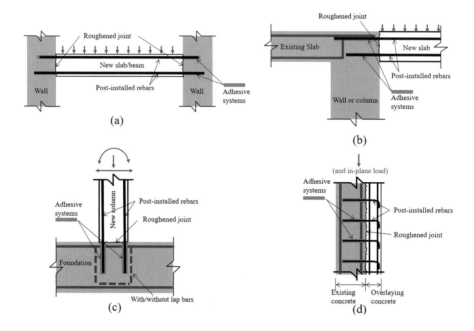

Figure 1.1: Typical application examples of post-installed reinforcements: (a) end anchor of new slab/beam attached onto walls (shear or diaphragm wall); (b) lap splice of new slab attached to existing slab; (c) end anchor with/without lap splice as moment resisting connection; and (d) new concrete overlays (e.g., for wall strengthening, column jacketing, and slab thickening) (note: no transverse reinforcement is used)

(2014)) and lap splicing length (i.e., Cl. 8.7 in EN 1992-1-1 (2004) and Chapter 25 of ACI 318 (2014)). On the contrary, the anchor procedure is provided in EN 1992-4 (2018) and Chapter 17 of ACI 318 (2014). More detailed discussions can be found in Charney et al. (2013) and Morgan (2015).

The load-slip performance of post-installed reinforcements under static loading installed by using a qualified system can be similar or exceed that of a cast-in reinforcement according to extensive research carried out by Spieth (2002), whilst Simons (2007) investigated the seismic behaviour. Thus, the design provisions of end anchorages for cast-in reinforcements can be extended to post-installed reinforcements along with qualified products. Specific guidelines have been produced that qualify post-installed reinforcements designed by using the reinforcement anchorage (RA) design procedure, for example, EAD 330087-00-0601 (2018) (which replaces the TR 023 (2006) published by the European Organisation for Technical Assessment (EOTA)) and the AC 308 (2016) in Europe and the United States (US), respectively. On the contrary, post-installed reinforcements that meet guidelines such as EAD 330499-00-0601 (2017) and AC 308 (2016) in Europe and the US, respectively, can be designed based on the bonded anchor (BA) design procedure. Table 1.1 is a summarised list of the most relevant international documents and their requirements in using post-installed reinforcements (with relevant documents on post-installed anchors), for ease of reference and to facilitate the discussions hereafter. Note that this document does not consider other standards (e.g., EN 1504-6:2006) that allow products to be certified by following different requirements, that do not address the most critical installation conditions, and are not tied to available design building codes.

Table 1.1: List of international documents for qualification and design of post-installed reinforcements

Document	Organisation	Role and function	Remarks
Qualification			
EAD 330087-00-0601 (2018)	EOTA	Qualification of post-installed reinforcements in Europe under static loading and fire exposure.	Replaces TR 023 (2006). Design in accordance with EN 1992-1-1 and 1992-1-2 (2004).
EAD 331522-00-0601-00-0601 (endorsed draft 2018)	EOTA	Qualification of post-installed rebars with mortar under seismic conditions.	Listed on EOTA website; pending publication in the Official Journal of the European Union* Design in accordance with EN 1992-1 (2004).
AC 308 (2016)	International Code Council Evaluation Service, Inc. (ICC-ES)	Qualification of post-installed reinforcements and adhesion anchors under static and seismic loading.	Use with test criterion to supplement ACI 355-4 (2011). Design in accordance with Ch. 25 of ACI 318 (2014)
ACI 355.4 (2011)	ACI	Qualification of post-installed adhesion anchors under static and seismic loading.	Design in accordance with Ch. 17 of ACI 318 (2014).
EAD 330499-00-0601 (2017)	EOTA	Qualification of post-installed anchors under static loading in Europe.	Design in accordance with EN 1992-4 (2018) or CEN/TS 1992-4-5 (2009)
EOTA TR 049 (2016)	EOTA	Qualification of post-installed anchors under seismic loading in Europe.	Design in accordance with EN 1992-4 (2018) or EOTA TR 045 (2013)
Design			
EN 1992-1-1 (2004)	CEN	General reinforced concrete design in Europe.	Design provisions for anchorage and splice length in Chapter 8.

Table 1.1: Cont'd

Document	Organisation	Role and function	Remarks
ACI 318 (2014)	ACI	General reinforced concrete design in the US.	Design provisions for development length (rebar design procedures) in Chapter 25, and anchor design procedures in Chapter 17.
EOTA TR 045 (2013)	EOTA	Guidelines for design of post-installed anchors in Europe.	Superseded by EN 1992-4 (2018)
EN 1992-4 (2018)	CEN	Standard for design of post-installed anchors in Europe.	
BS 8539 (2012)	BSI	Selection and installation of post-installed anchors in the UK.	Recommendations for anchors without European Technical Approval (ETA) qualification.

Note: * The first citation of EADs needs to be in the Official Journal of the European Union and subsequent to publication, and available for download on the European Commission website.

1.1.2 Relevance to the local Hong Kong design guide

In Hong Kong, the Buildings Department (BD) of the Hong Kong government recommends the use of the Code of Practice for Structural Use of Concrete 2013 (hereinafter HKBD 2013) for designing, constructing, and controlling the quality of RC buildings and structures where the concrete is made with normal weight aggregates. HKBD 2013 was drafted with substantial reference to the now defunct British standard BS 8110 Part 1 1997 (superseded by EN 1992-1-1:2004). The minimum end anchorage bond length requirement is similar in the two codes, where yield strength of the reinforcement is assumed. The clauses are Cl. 8.4 in HKBD 2013 and Cl. 3.12.8.3 in BS 8110 (1997). There are no specific calculations required for the lap length in HKBD 2013, but provisions are given based on some deemed-to-comply practices, commonly for the length of bars of different sizes. Compared to the requirements in EN 1992-1-1 (2004) (i.e., anchorage length in Cl. 8.4.4 and splicing length in Cl. 8.7.3) or the splicing development length formula in Cl. 25.4.2.3 of ACI 318 (2014), the anchorage or splice length design in HKBD 2013 has not accounted for different variables such as the shape/size of the bars, concrete minimum cover, confinement effects, casting position, and type of grout or epoxy used.

A review of the relevant clauses in HKBD 2013 showed that this code might not be directly applicable to post-installed reinforcements for the following reasons:

(1) Adhesion systems are not codified in the local codes of practice. The Hong Kong Building Authority has provided guidelines for approving post-installed anchors, but not post-installed reinforcements. Cementitious/polymer-based grout

dowels/reinforcing bars can be provisionally approved in construction based on a BD approval letter.

(2) The approach for reinforcement detailing at the joints in the local codes of practice often requires a long anchorage length, which makes post-installed reinforcements impractical, as bending of the bars is not possible. This is due to the conservative assumption that the design stress of the reinforcement reaches its yield strength.

Similar challenges are found with the use of generic RC design codes such as EN 1992-1-1 (2004) and ACI 318 (2014). Hence, the publication of the EAD 330087-00-0601 (2018) (which supersedes TR 023 (2006)), and AC 308 (2016) by the EOTA in Europe and the American Concrete Institute (ACI) in the US, respectively, is to qualify the use of post-installed reinforcements in concrete structures. Therefore, guidelines for post-installed reinforcements that refer to the latest technologies and correspond with HKBD 2013 are necessary for local practices in Hong Kong.

The aim of the Guide is to establish guidelines for installing, designing and assessing post-installed reinforcements that are subjected to mainly static loads (with an introduction on exposure to fire and seismic conditions) by referring to HKBD 2013, EN 1992-1-1 (2004), ACI 318 (2014), EN 1992-4 (2018), EAD 330087-00-0601 (2018), EAD 331522-00-0601 (endorsed draft 2018) and AC 308 (2016).

The Guide is to be used in conjunction with HKBD 2013 or other relevant design codes to design other RC elements or inspect existing structural elements that are not included in the Guide.

1.2   Post-installed and cast-in reinforcements

This section provides an introduction on several concepts that have importance to the readers of the Guide for differentiating the behaviour (including load transfer and failure process) between post-installed and cast-in reinforcements.

1.2.1 Local load transfer mechanism among reinforcements, bonding agent, and concrete

The bond strength of reinforcements is important for effectively transferring load to the surrounding concrete. It is commonly assumed for design purposes (uniform-bond model) that the distribution of the average bond stress along the embedded length of the reinforcement is constant for both cast-in and post-installed reinforcements.

The bond force is the force that moves a reinforcing bar along the length of its longitudinal axis with respect to the surrounding concrete. The bond strength is the maximum bond stress that can be sustained by a bar in concrete (ACI 408R, 2003). Figure 1.2 schematically shows the load transfer mechanism of cast-in and post-installed reinforcements under tension. In Figure 1.2(a), which shows the cast-in reinforcement, the load is mainly transferred by the mechanical interlocking of the ribs at the reinforcement-concrete interface. The reaction forces within the concrete are assumed to be in the form of compressive struts, which are inclined to the axis of the reinforcement. The vector bearing forces can be decomposed in the parallel and perpendicular directions to the longitudinal axis of the reinforcement. The sum of the parallel components is the bond force, whereas the radial components in the perpendicular direction cause circumferential tensile stresses in the surrounding concrete. If the concrete cover is 'small', the radial component forces may cause splitting cracks in the concrete. If the concrete cover is 'large', then pull-out or tensile failure of the reinforcement may occur. However, concrete cone failure is not taken into account in RC standards, as concrete is not supposed to bear tensile forces. The tensile forces of reinforcements are thus transferred through local struts (i.e., splices) or global struts (i.e., idealised strut and tie models).

For the post-installed reinforcement shown in Figure 1.2(b), the load transfer mechanism involves two steps. First, the load from the reinforcement is inhibited by the surrounding adhesive, which is similar to the load transfer mechanism of the cast-in reinforcements that are directly anchored in the concrete. Then, this load is transferred from the surrounding adhesive to the surrounding concrete through both adhesion and low friction. The adhesive layer is then subjected to lateral dilation from bearing the stress, which adds to the friction. The concrete subsequently develops hoop stresses around the reinforcement, which along with a small concrete cover, may lead to a splitting failure. Additionally, there may be failure from pull-out between the reinforcement and mortar or mortar and concrete, but the likelihood of failure is contingent on the strength of the individual layers. If the embedment is deep enough, then steel failure is also a possibility.

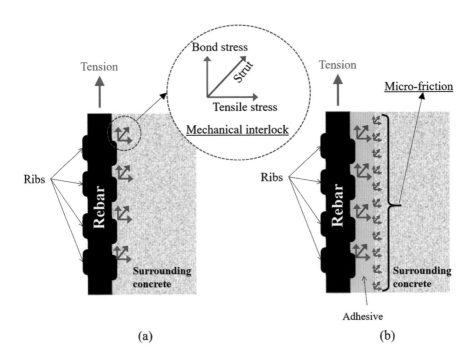

(a)   (b)

Figure 1.2: Local load transfer mechanism of (a) cast-in, and (b) post-installed reinforcements

1.2.2 Load transfer with reinforcement lap splicing and reinforcement anchoring and associated failure modes

The presence of an existing reinforcement adjacent to a new post-installed reinforcement under tension (similar to lap splicing with cast-in reinforcements) provides support for additional load transfer aside from the local load transfer between the reinforcements and concrete (or bonding agent), as discussed in the previous section. Figure 1.3(a) shows the load transfer mechanism through compressive struts in the concrete between two adjacent lapping bars. The resolved forces perpendicular to the longitudinal direction of the bars act as splitting forces, which are normally inhibited by the transverse reinforcements, with a negligible amount of the splitting forces attributed to the tensile capacity of the concrete. If the development length of the bar is not properly designed, splitting concrete failure, particularly at the near-edge bars, is highly possible. On the contrary, when there is no lap splicing, the splitting forces will be directly transferred to the concrete, which could result in typical anchorage failures, including steel failure, pull-out, concrete breakout, or bond and splitting failure (see Figure 1.3(b)).

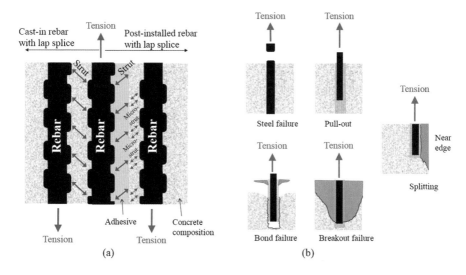

Figure 1.3: Load transfer mechanism of (a) lap splices in cast-in and post-installed reinforcements, and (b) without lap splices, anchor-dominated failure modes

1.2.3 Shear stress transfer at interface between new and old concrete

In concrete structures with post-installed reinforcements, the interface between new and old concrete should be adequately roughened, and the new concrete will bond to the old concrete (see Figure 1.4). At the end of the beam, the shear force transfers the tension stress to the bottom reinforcement through struts and ties. The concrete struts prevent concrete cone failure. In practice, the top reinforcement bars of simply supported beams may be in tension due to partial fixity. In such case, the top reinforcement should be examined for possible cone failure in the absence of confinement provided by the concrete struts. In any case, the general principles of the RC procedure assume that the reinforcing bars do not significantly bear the shear forces, although it has been noted that dowel action is attributed to the presence of longitudinal bars. Readers are encouraged to refer to some of the more established references on examples of strut-and-tie model (STM) designs, e.g., fib Bulletin 61 (fédération internationale du béton, 2011b).

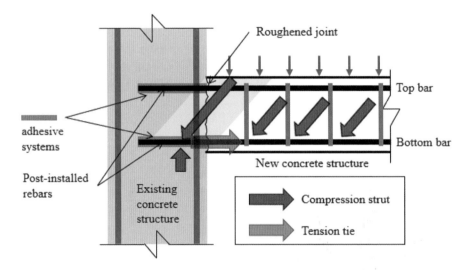

Figure 1.4: Strut-and-tie model of typical simply supported RC beam connected with post-installed reinforcements to existing RC structure

2

Adhesion Systems, Base Materials, and Reinforcements

2.1 Adhesion systems

2.1.1 Types of adhesives

There are three typical types of structural adhesives that are available for use in post-installed reinforcement systems, which include polymer adhesives, cementitious grouts, and hybrid grouts (which are a combination of polymers and cementitious components).

(1) Polymer adhesives
Polymer adhesives are usually two-component systems that can be delivered in the form of: (a) injection systems in a soft or hard plastic cartridge, or (b) capsule anchoring systems in a soft or glass capsule. The polymer adhesives used in injection systems are commonly pre-packaged in dual-plastic cartridges. The two products are then mixed in a static mixer before being dispensed through injection into a drilled hole (see Figure 2.1(a) on p. 10). Capsule anchoring systems encapsulate the adhesive in a glass, plastic, or foil capsule which is more robust and flexible for use in construction sites (see Figure 2.1(b)). The capsule is inserted into a drilled hole and then mixed when direct boring through the capsule is carried out to facilitate the setting of the reinforcement.

Polymer adhesives provide a higher bonding strength to old concrete. This is evident for post-installed reinforcements, as confined pull-out tests (without splitting failure, and in accordance with the testing condition specified in EAD 330087-00-0601 (2018)) have demonstrated that polymer adhesives can offer an average bond strength that ranges from 15 MPa to 35 MPa in uncracked low strength concrete for an embedment length of approximately 7 to 10 bar diameters. On the contrary, the average bond strength for cast-in reinforcements is approximately 10 MPa. Polymer adhesives have high compression strength that ranges from 50 to 100 MPa and high tensile strength that is usually between 10 and 40 MPa. Many of the polymer adhesives have adequate viscosity so that there are no voids in the bond layer if they are installed with the proper delivery system (see Chapters 3 and 5). These adhesives have excellent bonding strength, which allows installation in many different directions (i.e., vertically downwards or upwards, horizontally and inclined) in a variety of different use conditions.

Despite the advantages offered by polymer adhesives, there are some known setbacks. Polymer adhesives have a different coefficient of thermal expansion than the

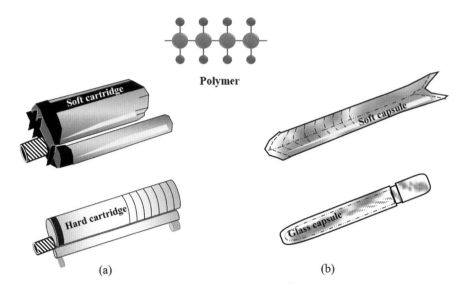

Figure 2.1: Types of polymer adhesives: (a) injection systems with soft or hard plastic cartridge, and (b) capsule systems with soft or glass capsule

concrete base material. The obvious difference in the modulus of elasticity between these adhesives (usually has a higher modulus of elasticity) and that of the concrete base material (usually has a lower modulus of elasticity) may induce cracks due to internal stresses. Nevertheless, the problem can be circumvented through a qualified system (see Chapter 3); for example, by using aggregate sieve grading to optimise the proportion of polymer by reducing the amount of polymer, thus reducing the differences in their natural properties (El-Reedy, 2008).

(2) Cementitious grouts
A typical cementitious grout is schematically shown in Figure 2.2. Whilst polymer adhesives offer physical protection to the steel reinforcements, cementitious grouts provide passive protection as they have the same properties as old concrete. The alkaline environment around the reinforcement is increased with cementitious grouts, thus protecting steel reinforcements against corrosion. This advantage of cementitious grouts means that they are often used for corrosion repair. However, a larger diameter of the drilled hole is necessary for the installation of cementitious grouts compared to other types of bonding materials. Low viscosity grout mixes are typically limited to vertical down-holes and not horizontal or overhead applications.

(3) Hybrid grouts
Hybrid adhesives are the combined products of polymer and cementitious grout. Liquid or powder polymers are added to the cement mortar to improve some of the properties of cementitious grout, particularly to increase its flexural resistance and elongation, reduce its water permeability, enhance the cementitious bond between new and old concrete, and increase the effectiveness of its performance. See Figure 2.3 on p. 12.

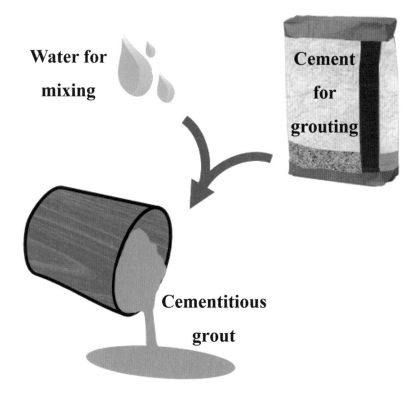

Figure 2.2: Cementitious grout (pouring)

2.1.2 Temperature effects on adhesives

It is common knowledge that adhesives are sensitive to temperature, and therefore the effects of temperature must be considered before, during, and after installation to ensure an effective connection. The main impacts of temperature on adhesives are commonly found in the following three scenarios (Gamache, 2017):

1. During storage of the adhesive, as the storage temperature can affect its shelf life;
2. At the time of installation, as both low and high temperatures of the concrete and adhesive have impacts on the gelling, curing time, and viscosity of the adhesive; and
3. During the service life of reinforcement structures, as high concrete temperatures can markedly affect the bond strength of post-installed reinforcement structures.

Special provisions should also be made for adhesives exposed to extremely high temperatures, such as fire.

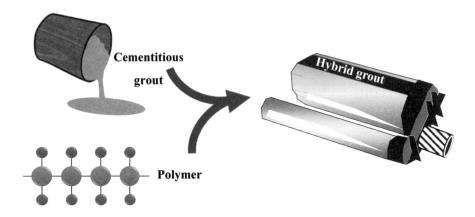

Figure 2.3: Hybrid grout

(1) Storage temperature

The storage temperature of adhesives prior to installation is commonly provided in the manufacturer's published installation instructions (MPII). The temperature limit should always be followed to prevent deterioration of the adhesive prior to installation. Special consideration should be given to the storage of adhesives at job sites (i.e., in lockboxes that are exposed to high temperatures under the sun). High temperatures may cause liquid separation and permanent deterioration of the adhesive, whist low temperature conditions or freezing may cause crystallisation. The bond strength of post-installed reinforcement structures may be greatly reduced due to the deterioration and crystallisation of the adhesive.

(2) Installation temperature

Adhesives undergo two processes (i.e., gelling and curing) when applied during the installation of post-installed reinforcements. The gelling time is basically its working time, which starts when dual adhesives in a carrier are mixed, thus initiating a chemical reaction. In a static mixing nozzle system, the gelling starts in the nozzle. During gelling, the adhesive can be manipulated, and the reinforcements can be adjusted, without affecting the later in-service strength of the post-installed reinforcement system. After the gelling time is completed, the adhesive (including the steel reinforcement element) must remain undisturbed until the curing time has elapsed. The curing time is the amount of time required for the mixed adhesive to reach full strength before being subjected to the desired load.

A general trend of the gelling and curing times versus temperature shows that longer gelling and curing times are caused by low temperatures, and vice versa, shorter gelling and curing times are due to high temperatures. Note that the gelling and curing times of adhesives depend on the temperature of the concrete base material (not the ambient air temperature) during installation. A low base material temperature may inhibit the curing process (longer gelling and curing times) and increase the viscosity of the adhesive, and thus it is less manageable during injection.

On the contrary, a high base material temperature may considerably reduce the gelling and curing times of the adhesive and reduce its viscosity, which would result in inadequate adhesive when carrying out horizontal and overhead installations. The high temperature environment of the base material causes rapid curing, which affects the use of post-installed reinforcements with deep embedment depths and large drilled holes.

A non-universal but fairly effective solution is to condition the adhesive to a temperature of approximately 20°C prior to use, which would allow optimum injection conditions. This measure is, however, ineffective in influencing the curing time, which is mostly related to the temperature of the concrete member, where the reinforcing bar is installed.

(3) Service temperature
The design of post-installed reinforcements, which includes the type of adhesive used, should cater to the service temperature of the base material (not the air temperature). The service temperature includes minimum, maximum, and extreme temperatures with short and long-term exposure of the reinforcement structure. In general, a high base material temperature reduces the bonding strength of the adhesive. On the contrary, low temperatures have little negative impact on the bonding strength, which has been verified in freeze-thaw tests (Gamache, 2017).

(4) Fire
For post-installed reinforcement systems that are subjected to high temperatures or fires, specific recommendations and details should be sought from the product manufacturer. Information on the variations in the adhesive bond strength with temperature obtained through different tests (see EAD 330087-00-0601 (2018) or other European national approvals) should be provided by the product manufacturer. Note that the temperature at different depths in the concrete will often be much lower than that at the concrete surface. Therefore, a longer embedment depth to compensate for the loss of bonding strength close to the surface of the concrete is beneficial for mitigating the effects of fire (IAN 104/15 (2015) and BS 8539: 2012).

2.1.3 Adhesion system use considerations

The use of an adhesive depends on its application, loading direction, environmental considerations, anchorage length, reinforcement diameter, drilling method, and job site conditions. Note that some anchoring adhesives should not be used for post-installed reinforcements. Only those that have been approved by EAD 330087-00-0601 (2018) (which supersedes TR 023 (2006)) or AC 308 (2016) in Table 3.8 are suitable.

The self-guiding questionnaire checklist in Table 2.1 is provided by the Guide to facilitate the selection of a suitable adhesion system. Engineers can first complete this questionnaire and then forward to the adhesive manufacturer or distributor for recommendations on the most appropriate product.

Table 2.1: Self-guiding questionnaire checklist to select adhesion system

1	What are the application conditions of the drilled hole? A. Dry B. Wet C. Water-filled/flooded*
2	What is the diameter and depth of the drilled hole? Rebar diameter: _____ Depth of hole: _____ *#
3	What is the direction of the drilled hole? A. Downwards B. Horizontal C. Overhead
4	What is the drilling method? _____
5	Does the concrete have any cracks? A. Yes B. No
6	What is the concrete cylinder/cube strength? _____/_____ MPa
7	What is the type of load imposed onto the post-installed reinforcement structures? A. Static load B. Quasi-static load C. Fatigue load D. Dynamic or seismic load E. Wind load.
8	Does the adhesive need to be fire resistant? A. Yes (_____ hours) B. No
9	What is the chloride content in the concrete? ^ _____%
10	What are the expected gelling and curing times? Gelling time: _____ Curing time: _____
11	What is the expected service life of the adhesive? _____ years
12	Is shrinkage of the adhesive a concern? A. Yes B. No
13	Storage temperature of adhesive: _____ Working temperature of concrete: _____ Service temperature range: from _____ to _____ Short-term^a temperature: from _____ to _____ Long-term^b temperature: from _____ to _____
14	Hole cleaning: A. Based on MPII B. Cannot clean thoroughly due to site conditions
15	Is the adhesive resistant to the following chemical products or environmental factors? (1) Alkaline products: drilling dust slurry pH = 12.6 or potassium hydroxide solution (10%) pH = 14. (2) Acids: acetic acid (10%), nitric acid (10%), hydrochloric acid (10%), or sulfuric acid (10%). (3) Solvents: benzyl alcohol, ethanol, ethyl acetate, methyl ethyl keton (MEK), trichlor ethylene, or xylol (mixture). (4) Products on job site: concrete plasticizer, diesel, engine oil, petrol, or oil from work. (5) Environment: saltwater; de-mineralised water, or sulphurous atmosphere (80 cycles)
16	Does the adhesive system need to meet ETA guidelines (EOTA EAD 330087, 2018) or ICC-ES guidelines (AC 308, 2016)? A. ETA guidelines B. ICC-ES guidelines

Notes:

* Condition is not taken into consideration in EAD 330087-00-0601 (2018) and AC 308 (2016); additional technical data from supplier is needed.

To be justified by engineer

^ Standards for chloride content tests can be obtained in BS 1881-124:2015 and EN 14629:2007.

a Short-term temperature: where higher concrete temperatures are temporary or part of a regular cycle of heating and cooling, such as day-night temperature rise and fall.

b Long-term temperature: where concrete temperatures may remain high over weeks or months.

2.2   Concrete

Several types of base materials can be used to accommodate post-installed reinforcement systems; for example, normal weight RC (cracked or non-cracked concrete), prestressed concrete, lightweight concrete, and masonry structures. In the Guide, only normal weight RC structures are elaborated as they are the most commonly used in Hong Kong.

The Guide applies post-installed reinforcements in normal weight structural concrete C12/C15 to C50/C60 (characteristic cylinder/cube strength in MPa) in accordance with EAD 330087-00-0601 (2018) and AC 308 (2016). For higher grade concrete (>70 MPa), which is permitted in RC designs in accordance with HKBD 2013, the bond strength of post-installed reinforcements should not exceed the value set for C60, unless the supplier provides technical data that supports its use. The minimum thickness of the concrete members in which the reinforcement will be installed should be greater than or equal to the sum of the minimum anchorage length of the post-installed reinforcement and the minimum cover thickness (refer to discussions in Chapter 4).

2.3   Reinforcements

Similar to concrete, steel reinforcements in this Guide should conform to the requirements of HKBD 2013, with the exception of plain steel reinforcements of Grade 250 which are not allowed to be used in post-installed reinforcement systems. That is, deformed carbon steel bars Grades 500B and 500C with a surface geometry (i.e., rib parameters, relative rib areas, longitudinal and transverse ribs) that complies to CS2:2012 shall be used for reinforcements. Note that the requirements of CS2:2012 for reinforcement geometry are the same as those in Annex C of EN 1992-1-1:2004.

3
Installation Methods

3.1 General

3.1.1 Installer qualifications

Post-installed reinforcements are to be installed by qualified technicians in accordance with construction specifications and, where applicable, the manufacturer's instructions (see Annexes F.2(b) and F.2(c) in EN 1992-4 (2018) and Cl. 17.8.1 in ACI 318 (2014)). The construction specifications must indicate that the installation of post-installed reinforcements is carried out based on the MPII. The manufacturer's recommendations for the specified adhesion system take precedence over all other recommendations.

Qualified adhesive anchor installers would have completed a recognised certification programme, as per Cl. R17.8.2.3 in ACI 318 (2014) which states: 'an equivalent certified installer program should test the adhesive anchor installer's knowledge and skill by an objectively fair and unbiased administration and grading of a written and performance exam. Programs should reflect the knowledge and skill required to install available commercial anchor systems. The effectiveness of a written exam should be verified through statistical analysis of the questions and answers. An equivalent program should provide a responsive and accurate mechanism to verify credentials, which are renewed on a periodic basis'. A similar certification programme should also be implemented for qualifying the installers of post-installed reinforcements.

The installer is also required to be trained and supervised by a competent trainer based on BS 8539 (2012) for the following:

(1) On-site training
The trainer is to provide on-site training to the installer on the correct setting of the tools used to properly install post-installed reinforcements and how to carry out the drilling process. Note that training from other projects would not be considered adequate, as installation requirements vary depending on the type of adhesive used.

(2) Knowledge
The trainer is to inform the installer about the functions of post-installed reinforcements, and highlight the consequences of non-compliance with the installation procedures provided in the MPII and material safety data sheets.

(3) Experience

The trainer is to provide closer supervision if the installer is less experienced (see Chapter 5).

3.1.2 Installation process

The performance of post-installed reinforcements depends on their installation procedure. A typical sequential procedure for installation is provided below and illustrated in Figure 3.1.

(1) Determine the location of existing reinforcements.
(2) Roughen the surface of the old concrete.
(3) Use a suitable drilling machine to drill a hole into the concrete to the required depth with the correct nominal diameter drill bit.
(4) Clean the drilled hole in accordance with the MPII.
(5) Inject adhesive into the drilled hole in accordance with the MPII.
(6) Set the reinforcement to the required embedment depth before the working time has elapsed in accordance with the MPII.

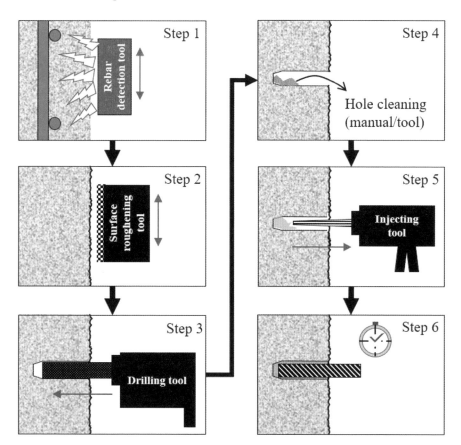

Figure 3.1: Typical sequential procedure for installing post-installed rebar

3.2   Locating existing reinforcements (Step 1)

Locating existing reinforcements embedded in old concrete prior to drilling is an important step to avoid interference with existing reinforcements, which can be done by using different scanners, depending on the following situations:

(1) Magnetic field scanners—suitable for locating reinforcing bars embedded within 200–250 mm of the concrete surface;
(2) Ground-penetrating radar (GPR) scanners—suitable for ferrous or non-ferrous embedded items (e.g., aluminium); and
(3) X-ray scanners—suitable for congested or deeply embedded reinforcements that cannot be detected with magnetic field or GPR scanners.

All of the mentioned scanners are best supplemented with as-built drawings or design documents, which can be accessed through digital programmes installed on electronic devices such as tablets or laptops. One of the widely used digital programmes is Building Information Modelling (BIM), which is further enhanced with the use of augmented reality.

3.3   Roughening old concrete surface (Step 2)

3.3.1 Roughened area

Roughening the surface of old concrete prior to the casting of adjoining fresh concrete can enhance adhesion and joint friction. The carbonated layer should be removed in areas that will receive the post-installed reinforcing bars. In general, this comprises a circular area with a diameter (d_{rough}) that equals the diameter of the bar (ϕ) and an additional 60 mm (i.e., $d_{rough} = \phi + 60$ mm) (see Hilti, 2016) The requirements for roughening concrete surfaces found in HKBD 2013 and EN 1992-1-1 (2004) are briefly discussed in the following section.

3.3.2 Requirements in HKBD 2013

Cl. 10.3.10 in HKBD 2013 recommends that, similar to the roughening of a construction joint, roughening an old concrete surface can be done through applications of fine water spraying, stiff brushing, sand-blasting, or using a scale hammer. The concrete surface must be clean and free of loose particles. Roughening through applications of fine water spraying and/or stiff brushing should be done approximately 2 to 4 hours after the concrete is placed. Excessive roughening that damages or dislodges the coarse aggregate particles should be avoided.

3.3.3 Requirements in EN 1992-1-1 (2004)

Compared to the deemed-to-comply qualitative approach in HKBD 2013, Cl. 6.2.5 in EN 1992-1-1 (2004) provides a quantifiable equation for roughening (shear friction) in the design of post-installed reinforcing bars.

The shear stress demand (τ_{Edi}, defined in Eq. (3.1)) at the interface should be less than the shear stress capacity (τ_{Rdi}, defined in Eq. (3.2)).

$$\tau_{Edi} = \beta\, V_{Ed} / (z\, b_i) \qquad\qquad\qquad\qquad\qquad\qquad\qquad \text{Eq. (3.1)}$$

where:

β is the ratio of the longitudinal force in the new concrete area and the total longitudinal force in either the compression or tension zone, both calculated for the section considered;

V_{Ed} is the transverse shear force;

z is the lever arm of the section; and

b_i is the width of the interface.

$$\tau_{Rdi} = c\, f_{ctd} + \mu\, f_n + \rho\, f_{yd}\, (\mu \sin \alpha + \cos \alpha) \leq 0.5\, v\, f_{cd} \qquad\qquad \text{Eq. (3.2)}$$

where:

c and μ are factors that depend on the roughness of the interface;

f_{ctd} is the design value of the tensile stress capacity of the concrete;

f_n is the stress per unit area caused by the minimum external normal force across the interface that can act simultaneously with the shear force, positive for compression and negative for tension. When f_n is tension, c is equal to zero;

ρ is the ratio between the area of the reinforcement (A_s) that crosses the interface, including ordinary shear reinforcement and the area of the joint (A_i);

f_{yd} is the design value of the yield stress capacity of the steel reinforcement connector;

α is the inclination angle formed by the longitudinal axis of the reinforcement with the contact interface, and limited to between 45° and 90°;

v is a shear strength reduction factor for cracked concrete; and

f_{cd} is the design value of the cylinder strength capacity of the concrete.

A similar formulation, but more detailed variation of Eqs. (3.1) and (3.2), is given in Cl. 7.3.3.6 of the fib Model Code for Concrete Structures 2010 (fédération internationale du béton, 2013): an interaction coefficient (κ_1) for tensile force activated in the dowel bar and interaction coefficient (κ_2) for flexural resistance are provided. Structural designers should use Eq. (3.2) with caution, as the reinforcements are assumed to yield. Note that in post-installed reinforcement systems, steel yielding failure mode can only be achieved with longer anchorage length. However, longer anchorage lengths might not be plausible in some applications (e.g., jacketing of columns or thickening of slabs). A more detailed discussion on failure modes is provided in Chapter 4.

3.3.4 Methods of surface preparation

A comprehensive technical note on surface preparation can be found in Mailvaganam et al. (1998). The methods are summarised as follows:

(1) Chemical cleaning

Concrete surfaces contaminated with oil, grease, or dirt can be cleaned with chemicals (i.e., detergent, trisodium phosphate, or other concrete chemical cleaners/degreasers). Residuals that are left on the surface after the application of chemicals should be removed by scrubbing and rinsing with water. Refrain from using solvents, as they dissolve contaminants, which will seep further into the concrete. Also note that muriatic acid, which is a popular chemical agent used for a variety of cleaning purposes, is relatively ineffective for removing oil and grease.

(2) Mechanical cleaning

Mechanical cleaning removes impurities by using devices with impact tools, including bush hammers, scabblers, and needle guns, as well as rotary tools, such as sanding discs and grinding wheels, which can be effective for cleaning lower compressive strength concrete. Two mechanical cleaning methods, scabbling and scarifying, are elaborated below.

   (a) Scabbling

      Scabbling is done through impact tools such as bush hammers (see Figure 3.2), scabblers (which use compressed air to hammer piston-mounted bits into the surface of concrete; see Figure 3.3), and needle guns, which can effectively remove several millimetres of the concrete surface. Whilst scabbling can roughen the surface more so than blasting, the operations are nevertheless dusty, noisy, and produce vibration. It may be necessary to use water jetting or wet sandblasting for a final surface cleaning after the use of such impact tools.

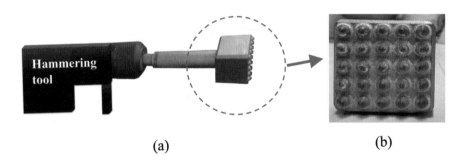

(a)   (b)

Figure 3.2: Example of a bushing tool: (a) general view, and (b) mechanical surface

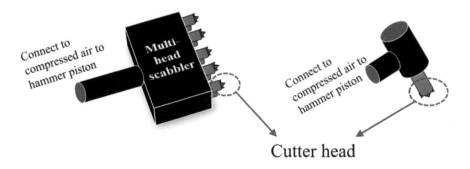

Figure 3.3: Schematic diagram of scabbler

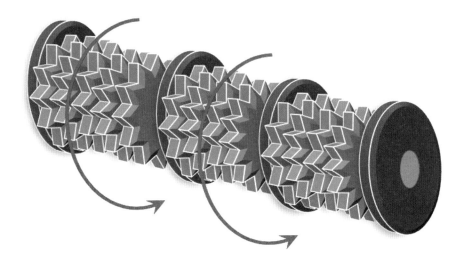

Figure 3.4: Schematic diagram of rotating circular wheel in a scarifying machine

(b) Scarifying

Scarifying is done with scarifiers or machines that are equipped with a rotating circular cutting wheel (see Figure 3.4). The depth of the cut can be more precisely controlled than with the use of scabblers. However, as is the case with scabblers, the use of scarifiers is a dusty and noisy process that produces vibrations, even though the amount of dust created by scarifiers can be controlled due to an attached dust collector. Nevertheless, scarifiers are relatively expensive and heavy, and require skilled operators.

(3) Blast cleaning

Blast cleaning can be done in the form of abrasive sand-blasting (both wet and dry), shot-blasting, or water-blasting.

(a) Sand-blasting

Sand-blasting is carried out with machines that eject a high velocity of sand (with a particle size that ranges from 2 mm to 2.4 mm) with compressed air from a nozzle. The distance between the nozzle and surface to be sand-blasted depends on the hardness of the concrete. Wet sand-blasting uses a high-powered stream of water with sand to remove impurities, while dry sandblasting uses high-pressure air and abrasive sand. However, dry sand-blasting often causes dust pollution and requires extensive protective measures during implementation.

(b) Shot-blasting

Shot-blasting is usually done by shooting a metallic abrasive (i.e., steel shot) to scour the concrete surface. The shots are propelled through a rotating blast wheel, which impact the concrete surface and then rebound into a separator along with the concrete debris. The debris is then separated from the shots.

Shot-blasting is commonly used to clean or scarify concrete up to a depth of 3 mm. The major advantage of shot-blasting is its ability to control dust.

(c) Water-blasting

During water-blasting, an adjustable high-velocity and high-pressure water jet is used to roughen the concrete surface. High velocity jets of water are propelled through a series of adjustable nozzles across a boom. Water is uniformly moved back and forth across the concrete surface as the water blasting unit moves forward. This method can be used to scarify concrete up to a depth of 30 mm. However, this method is mainly limited by the collection and daily disposal of wastewater, as well as the removal of debris on a daily basis to prevent the debris from hardening. Nevertheless, the advantages of this method outweigh its disadvantages, as there is no produced dust, little noise, no induced-vibration, selectively removes old concrete but not good concrete (as opposed to jack hammering), and does not damage the reinforcements (as opposed to scabbling and scarifying).

(4) Acid Etching

Whilst acid etching removes just the appropriate amount of cement paste for a roughened surface, there is unfortunately the potential of reinforcement corrosion. Therefore, acid etching is the least recommended method and should only be used when no other alternative means of cleaning the concrete surface is feasible.

(5) Flame cleaning

Flame cleaning is generally used to clean concrete surfaces that will be subsequently coated or overlaid with a resin finish. The process is carried out with a multi-flame oxy-acetylene blowpipe, which is passed over the concrete surface at a uniform speed of 0.02 m/s to 0.03 m/s. The surface is spalled and melted, which is why the moisture content of the concrete has the largest impact because dry concrete does not spall very much. However, concrete that is saturated before flame treatment would allow uniform removal to be achieved.

3.4   Drilling holes into concrete (Step 3)

3.4.1 Hole drilling requirements

Before drilling takes place, the installer should have good knowledge on drilling and must be equipped with adequate information on the drilling task itself (i.e., type of equipment required, the different metres of drill bits needed, the depth and direction of the drilling, the thickness of the concrete structure, edge distance, spacing between the drilled holes, and drilling aids for deep hole drilling) in order to meet the design requirements of the drilled holes. Drill bits recommended by the manufacturer should be used, and the depth of the holes should conform to the manufacturer's instructions. The size of the drilled hole is based on the ETA requirement applicable to the drill bit. Note that for large drill diameters, wet coring bits are usually used. However, wet coring bits produce a smoother internal surface of the drilled hole, which will reduce bonding.

3.4.2 Types of drilling methods

There are a variety of different drills on the market; for example, rotatory drills, impact drills, hammer drills, compressed air drills/percussive rock drills, core drills, magnetic drills, and electric drills. However, hammer, compressed air/percussive rock, and core drills will be discussed in view of the typically deeper embedment required to meet the design guidelines for longer anchorage lengths in some of the post-installed reinforcements. These three different types of drills are elaborated as follows:

(1) Hammer drills

Hammer drills are common, portable, and user-friendly. Hence, they are often the preferred tool for most applications. Figure 3.5(a) shows a typical hammer drill with solid and hollow drill bits. Hammer drills produce relatively rough and non-uniform holes, which should be thoroughly cleaned to enhance bonding. Even though they are common, hammer drills are not as practical for drilling longer holes and in cases for cutting through existing reinforcements when required. The dust produced when a hammer drill is used is its major setback. Nevertheless, hollow drill bits can be utilised for concurrent drilling and the holes can be cleaned by vacuum extracting the debris.

(2) Compressed air/percussive rock drills

Compressed air/percussive rock drills are usually more efficient and can drill faster (see Figure 3.5(b)). Similar to hammer drills, they produce a rough drilled hole which is suitable for bonding. Nevertheless, the larger impact energy may damage concrete with small edge distance or reduced backside cover, which means that these drills are more suited for concrete with larger edge distance/spacing/member thickness.

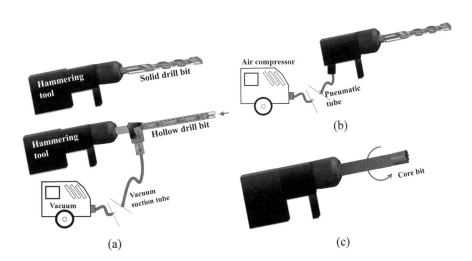

Figure 3.5: Drilling methods: (a) hammer drill with solid and hollow drill bits, (b) compressed air drill, and (c) core drill

(3) Core drills
Core drilling (dry or wet) is carried out with diamond core drill bits that abrade the concrete (see Figure 3.5(c)). With extensions, core drills can produce long and straight holes, and therefore are generally more preferred for deep embedment lengths. The hollow core bits have adequate stiffness to core in alignment with the intended path and are capable of drilling through existing reinforcements. Although core drills are highly efficient especially in the case of large drill diameters and deep embedments, care must be taken to avoid unintentional cutting of existing reinforcements. If there is interference with existing reinforcements during drilling, consultation with an engineer is required, who will then determine a new hole and evaluate the effects on the attachment. The smooth holes produced by core drills are usually covered with a thin film of dust, which is unfavourable for bonding. Hence, core drilled holes must be thoroughly cleaned prior to injecting the adhesive (see Section 3.5 for Step 4 on cleaning holes). Note that not all adhesives can provide a sufficient bonding strength in core drilled boreholes.

3.4.3 Drilling aids

A variety of drilling aids are usually required to ensure a perpendicular angle of the drilled holes with the concrete surface, depth of the drilled holes, and roughening of the hole surface.

(1) Drilling direction
In the drilling of deep holes (> 200 mm), the direction of the drilling could be supported by, for example, using an alignment tool (see Figure 3.6(a)), bubble level, or visual inspection check to ensure that the holes are perpendicular to the concrete surface.

(2) Drilling depth
The drilling depth can be managed with simple techniques such as using a piece of adhesive tape as a marker (see Figure 3.6(b)).

(3) Hole surface roughening
To address the challenges of diamond core drilled holes, which have a thin layer of accumulated dust on the newly formed surface (thus reducing bonding), roughening tools can be used to increase the roughness of the surface of the hole. Note that roughening tools can only be used in combination with products that meet specific qualified products (see qualification requirements in Chapter 5).

3.5   Cleaning drilled holes (Step 4)

Hole cleaning is essential to ensure that debris and dust do not reduce the strength of the bond between the injected adhesive and the surface of the hole. An example of the effects of the bond strength with different types of hole cleaning methods is shown schematically in Figure 3.7.

Hole cleaning depends on the site conditions and the limitations of the manufacturer's products. The task should be done in compliance with the MPII and the

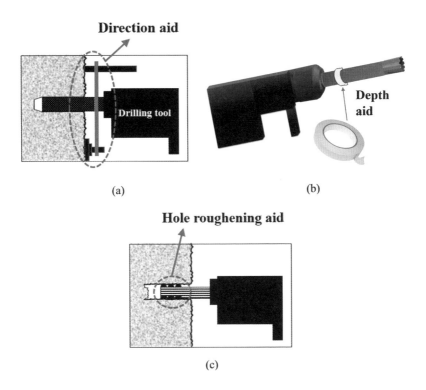

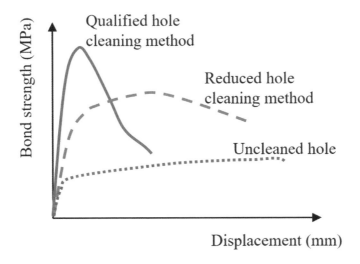

Figure 3.6: Drilling aids: (a) alignment tool for drilling direction, (b) adhesive tape marking for drilling depth, and (c) special hole surface roughening tool

Figure 3.7: Schematic representation of bond strength under the effects of different levels of cleaning

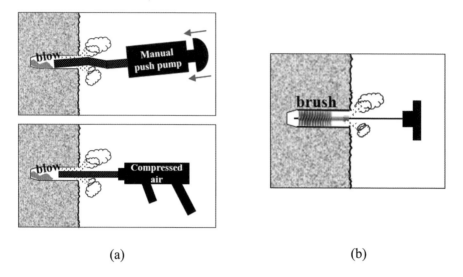

(a)   (b)

Figure 3.8: Accessories for hole cleaning: (a) manual or compressed air, and (b) mechanical brush

qualification standards for post-installed reinforcements, such as EAD 330087-00-0601 (2018) or AC 308 (2016) in Europe and the US, respectively. Hole cleaning generally involves three steps with specific repeated tasks: (i) repetitive cleaning of dust or debris by using air (manual push pump or compressed air (see Figure 3.8(a)), (ii) repetitive mechanical brushing of the hole wall to remove surface dust (see Figure 3.8(b)), and (iii) repeat Step (i) with compressed air.

To ensure worker safety, it is recommended that both a dust collector and eye protection are used. Additionally, the use of hollow drill bits that are connected to a vacuum system can greatly reduce the amount of dust produced. If wet core drilling is carried out, then the holes that have been flushed with water should be dried by applying compressed air.

3.6   Injecting adhesive (Step 5)

3.6.1 Inspection

The installer should inspect the product by referring to the MPII, and also consult the manufacturer before installing reinforcements in wet or flooded holes, as some adhesive materials cannot be applied in such conditions. If installation in wet or damp holes is permitted, the curing time must be increased due to the cooling effect of the water. Unless otherwise stated, the curing time has to be doubled for wet conditions.

3.6.2 Adhesive dispensing tools

There are generally three types of adhesive dispensing tools available in the market: (i) manual, (ii) pneumatic, and (iii) battery powered dispensers. The choice of dispenser largely depends on the volume of installations and reinforcement diameter.

3.6.3 Injection process

The formation of air voids during injection reduces the bond area, inhibits curing, causes reinforcement installation difficulties, and may cause uncontrollable ejection of the adhesive due to forced-out air. Hence, to ensure that the adhesive is void-free (or has the absolute minimum amount of entrained air), several precautionary measures are recommended for adoption:

(1) The adhesive components must be thoroughly mixed (i.e., completely homogeneous) before they are injected through the nozzles. In a new cartridge of adhesive, the starter portion must be discarded.

(2) The adhesive should be injected from the bottom of the hole until it is approximately two-thirds filled. For narrow or deep holes, extensions and piston plugs can be placed onto the nozzle (see Figure 3.9(a)). Piston plugs allow an almost air bubble free injection of the adhesive. For horizontal and overhead applications, the extensions and piston plugs can be equipped with a stopper or end cap during reinforcement installation (see Figure 3.9(b)).

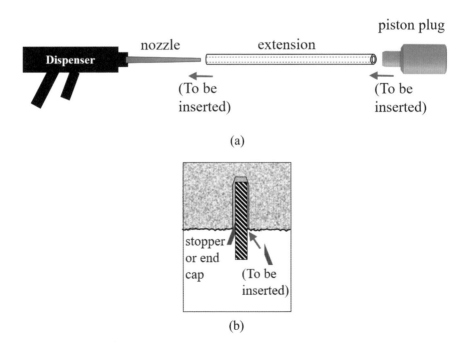

Figure 3.9: Accessories to reduce entrained air: (a) piston plug and its extensions, and (b) stopper and end caps for horizontal and overhead installations

During the injection process, the installer should pay attention to the gelling time and the final curing time based on the current temperature of the base material (refer to Section 2.1.2(2)). The installer should change the nozzle if the dwell time of the unused adhesive is more than the gelling time. For safety purposes, proper skin and eye protection must always be worn during the injection of adhesives.

3.7   Inserting reinforcements (Step 6)

3.7.1 Preparing reinforcements

The reinforcements to be installed into the drilled hole should be dry, free of oil, rust, or other residue and should not be too hot or cold. Embedment depth markers (e.g., adhesive tape) can be used for reference.

3.7.2 Inserting reinforcements

After injecting and filling two-thirds of a hole with adhesive, the reinforcement should be installed with a slow twisting motion as it is pushed downward to help distribute the adhesive around the reinforcement and eliminate trapped air (see Figure 3.10). In doing so, this would ensure that the entire area between the reinforcement and the surface of the holes is filled with adhesive.

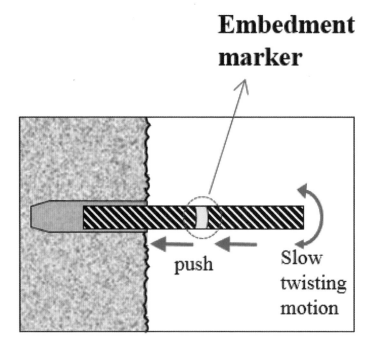

Figure 3.10: Rebar insertion

A proper installation should ensure that:

(1) the desired anchoring embedment is reached by referring to the embedment mark on the concrete surface,

(2) there is excess adhesive flow out from the borehole after the reinforcement has been fully inserted to the embedment mark. If adhesive overflow is not observed, the reinforcement must be removed and re-installed,

(3) the reinforcement is supported and secured in place until the adhesive starts to harden, especially for overhead reinforcement installations. Early contact or premature loading could partially destroy the bond between the concrete and adhesive. Full loads may only be applied after the curing time has elapsed, and

(4) the gelling or working time, and the final curing time must be observed, based on the current temperature of the base material.

The tendency of the reinforcement to spring back after being pushed into the adhesive mass points to the presence of air bubbles in the adhesive, or a popping sound can be heard as air bubbles are displaced upward. In this situation, the bar should be pulled out first and then the hole drilled and cleaned again after the curing time of the adhesive has elapsed.

Reinforcements with a smaller diameter can be inserted in a vertical downward direction with relatively minimal effort. On the contrary, large diameter reinforcements may require substantial effort to lift and case-specific methods are required to ensure that the installation is completed within the gelling time of the adhesive.

Installation of horizontal or upwardly inclined reinforcements, larger diameter reinforcements, and reinforcements that carry sustained tension loads should be conducted by a qualified installer under the ongoing supervision of an inspector. This inspector has to then submit a report to a licensed design professional and building official.

4
Design Methods and Examples

Different design methodologies have been developed in Europe and the US which allow for the use of post-installed reinforcing bars. This chapter discusses the available design methods and proposes a rational option for post-installed reinforcements in accordance with HKBD 2013. Written with substantial reference to the now defunct BS 8110 Part 1 1997, HKBD 2013 is compatible with the Eurocodes rather than the US codes. Hence, the European approach (i.e., the (i) RA design procedure: prequalification with EAD 330087-00-0601 (2018) then designed in accordance with EN 1992-1-1 (2004), and the (ii) BA design procedure: prequalification with EAD 330499-00-0601 (2017) then designed in accordance with EN 1992-4 (2018)) will be used as the background documents that provide a local design method accommodating the context of Hong Kong. Readers of the Guide are encouraged to refer to fib Bulletin 58 (fédération internationale du béton, 2011a) for additional reading material.

4.1 Identifying key design parameters

The design of post-installed reinforcing bar connections requires the engineer to determine the type of reinforcement, size, spacing, anchorage, and splice lengths of the reinforcement, as well as the quantity of reinforcements. The key parameters of the existing structure, site constraints, and arrangements of the connections which would affect the connection design are summarised in Table 4.1 on p. 31.

4.2 Design philosophy for post-installed reinforcements: Reinforcement and bonded anchors

4.2.1 Comparison of current provisions for post-installed reinforcement designs: European standards

The international documents listed in Table 1.1 are used to facilitate the discussion in this section. In general, post-installed reinforcements can be rationally designed based on the RA or BA design procedures, with differences in assumptions and being subjected to limitations. For the prequalification of post-installed reinforcements that use the RA design procedures, EAD 330087-00-0601 (2018) is used to determine the suitability of the adhesion system for post-installed reinforcements. After an adhesion

Table 4.1: Factors affecting post-installed rebar connection designs

Pre-qualification	(1) Adhesive assessment standard Rebar anchorage design procedures: EAD 330087-00-0601 (2018) or bonded anchor design procedures: EAD 330499-00-0601-00-0601 (2017)
Base materials and reinforcement	(1) Strength grade of concrete (2) Condition of concrete (cracked or uncracked, carbonated or non-carbonated, etc.) (3) Maximum chloride content in concrete (4) Ultimate bond strength and design bond strength of adhesive (5) Minimum thickness of base material (6) Yielding strength of reinforcement
Jobsite constraints	(1) The minimum and maximum concrete temperatures at time of installation and during the entire design life (2) Access and geometrical constraints on jobsite
Installation	(1) Requirements for preparation/roughening of existing concrete surface (2) Requirements for hole drilling (hammer, core, or compressed air drill) (3) Hole diameter (4) Orientation of connection (downward, horizontal, or overhead) (5) Environmental condition of concrete (dry, water-saturated, water filled, or flooded) (6) Existing reinforcement layout and size as given in drawing and confirmed on site with detection equipment (7) Requirements for training/certification of installers and supervisor
Design requirements	(1) Design code (rebar anchorage design procedures: EN 1992-1-1 (2004) and HKBD 2013 or bonded anchor design procedures: EN 1992-4 (2018) (2) Design life (3) Load type (sustained, static, quasi-static, seismic, shock, and wind) (4) Fire requirements (5) Corrosion resistance (6) Creep (7) Fatigue (8) Seismic

system has been suitably qualified, post-installed reinforcements can be designed using the RA design procedures based on Chapter 8 in EN 1992-1-1 (2004).

On the contrary, EAD 330499-00-0601 (2017) (formerly Part 5 of ETAG 001, 2013) provides provisions to determine the suitability of mortars or adhesives used for the anchor in the prequalification of bonded anchors. After an adhesion system has been suitably qualified, the anchor can then be designed in accordance with EN 1992-4 (2018). Table 4.2a provides a general comparison of both design methods.

Table 4.2a: Comparison of rebar anchorage and bonded anchor design procedures, using most relevant European standards

Main difference	Rebar anchorage design procedures	Bonded anchor design procedures
Adhesive assessment qualification documents	Under static conditions and fire exposure: EAD 330087-00-0601 (2018) Seismic conditions: EAD 331522-00-0601 (endorsed draft 2018)	Under static conditions: EAD 330499-00-0601 (2017) Under seismic conditions: TR 045 (2013)
Design standard	Under static conditions: Chapter 8 in EN 1992-1-1 (2004) Under seismic conditions: Chapter 5.6 in EN 1998-1 (2004)	EN 1992-4 (2018)
Load direction	Tension	Tension, shear, combination of both
Load transfer mechanism	Equilibrium with local or global concrete struts, may require the supplement of transverse reinforcement in lapping splices	Utilisation of tensile concrete strength
Failure mode	<u>Tension</u>: steel failure, pull-out, splitting (near the edge)	<u>Tension</u>: steel failure, cone-shaped concrete breakout (cone failure), bond failure (pull-out), splitting (near the edge) <u>Shear</u>: steel failure, cone-shaped concrete breakout, concrete pryout
Provision to base material	Uncracked concrete*	Cracked and uncracked concrete
Partial safety factor	$\gamma_s = 1.15$**	$\gamma_s = 1.2 \ (f_{uk}/f_{yk}) \geq 1.4$** (tension loading)
Basic design value of rebar	Yield strength of rebar	Ultimate strength for anchor and yield strength for rebar

Table 4.2a: Cont'd

Main difference	Rebar anchorage design procedures	Bonded anchor design procedures
Basic design value of bond strength	Deduced by calculation (associated with concrete tensile strength)	Tested and approved (associated with bond strength)
Design steps	a) Calculation of required steel cross-section of reinforcement b) Calculation of required embedment length	a) Calculation of all characteristic capacities b) Determination of minimum capacity that controls failure anchorage
Design results	Reinforcement length	Strength capacity
Allowable embedment length (l_b)	max $\{0.3\ l_{b,rqd};\ 10\ \phi;\ 100\ mm\}$ $\leq l_b \leq 60\ \phi$ (ϕ is the rebar diameter)	$6\ \phi \leq l_b \leq 20\ \phi$ (ϕ is the rebar diameter)

* The equivalence in terms of pull-out resistance in cracked concrete between a post-installed rebar system and a cast-in bar is checked in the qualification, as per EAD 330087-00-0601 (2018).

** For convenience of comparison with the local practice in Hong Kong, Table 4.2b provides more details on the partial safety factors for both the rebar anchorage and bonded anchor design procedures.

Table 4.2b: Comparison of partial safety factors for rebar anchorage and bonded anchor design procedures, as per European standards and Hong Kong standards and practices

Design procedures	Rebar anchorage	Bonded anchor	
Design standards/ practices	HKBD 2013 (at ULS)	EN 1992-4 (2018) (at ULS)	HK practice (global factor of safety)
Failure mode			
Steel	1.15	(a) $1.2\ (f_{uk}/f_{yk}) \geq 1.4$ (tension loading) (b) $1.0\ (f_{uk}/f_{yk}) \geq 1.25$ (shear loading, for $f_{uk} \leq 800$ N/mm^2 and $f_{uk}/f_{yk} \leq 0.8$) (c) $= 1.5$ (shear loading, for $f_{uk} > 800$ N/mm^2 and $f_{uk}/f_{yk} > 0.8$)	3.0
Concrete cone	N/A	≥ 1.5	3.0
Concrete edge (splitting)	N/A	≥ 1.5	3.0

4.2.2 Design provisions for RA design procedures in HKBD 2013

This section reviews the provisions for anchorage, detailing cast-in straight longitudinal reinforcements in a local code, HKBD 2013, and an international code, EN 1992-1-1 (2004), which use the RA design procedures for post-installed reinforcements. The procedure and equations used for the calculation of the anchorage and lap lengths are also elaborated.

Straight bar anchorage (Cl. 8.4)

Eq. (4.1) shows the derivation of the basic anchorage length (l_b) with the assumption of force equilibrium and considers that the resistance of the anchorage bond force (F_{bond}) is greater than the compression or tension force in the steel bar (F_{rebar}).

$$F_{bond} \geq F_{rebar}$$

$$f_{bu} A_{s,surface} \geq f_{rebar} A_s$$

$$f_{bu}(\pi\phi)l_b \geq f_{rebar}(\pi\phi^2/4)$$

$$l_b \geq \frac{f_{rebar}}{f_{bu}}\frac{\phi}{4} \qquad\qquad \text{Eq. (4.1)}$$

where f_{bu} is the factored bond stress capacity, $A_{s,surface}$ is the lateral surface area of the steel bar bonded with the concrete base material, f_{rebar} is the stress (tensile or compression) in the reinforcement, A_s is the cross-sectional area of the reinforcement, and ϕ is the reinforcement diameter.

According to Cl. 8.4.5 of HKBD 2013, the bar is assumed to be fully stressed to its design yield strength ($0.87f_{yk}$) for the start of the anchorage length. Hence, Eq. (4.1) yields Eq. (4.2), which incorporates a material partial safety factor (γ_m) in the design (i.e., 0.87, which is the reciprocal of $\gamma_m = 1.15$).

$$l_b \geq \frac{0.87 f_{yk}}{f_{bu}}\frac{\phi}{4} \qquad\qquad \text{Eq. (4.2)}$$

The factored bond stress capacity (f_{bu}) according to Cl. 8.4.4 is a function of the concrete characteristic cube strength ($f_{cu,k}$). Eq. (4.3) shows the bond stress estimation with a coefficient β to implicitly account for the reinforcement type and stresses.

$$f_{bu} = \beta\sqrt{f_{cu,k}} \qquad\qquad \text{Eq. (4.3)}$$

where, for common ribbed bars, β is 0.50 and 0.63 for tension and compression stresses, respectively. This value includes a partial safety factor for bond stress (γ_m) of 1.4.

Lapped splices (Cl. 8.7.3)

There might be a situation where post-installed reinforcing bars are used to lap existing cast-in bars. Eq. (4.4) summarises the minimum required lap length ($l_{o,min}$).

$$l_{o,min} \geq \max\{15\phi, 300 \text{ mm}\} \qquad\qquad \text{Eq. (4.4)}$$

Special consideration is required for lap splicing in tension, as the location (i.e., top, bottom, or corner of a section) and concrete cover must be taken into account to decide on a factor of 1.4 or 2.0 times the minimum lap length. For lap splicing in compression, the factor is 1.25 times the minimum lap length.

Simplified rules for simply supported beams (Cl. 9.2.1.5 and Cl. 9.2.1.7)

Reinforcement that resists at least 15% of the maximum mid-span moment is to be used as the top bars to resist the negative moment that develops at the support due to partial fixity, despite the assumption of simply supported ends. Fifty percent (50%) of the calculated mid-span bottom reinforcement is to be provided as bottom bars near the support in a simply supported beam. Eq. (4.5a) shows the detailing requirement of the straight anchorage length for simply supported beams. Bends and hooks are not addressed as they are irrelevant to post-installed reinforcements.

$$l_{beam,sr} = \{12\phi \text{ after centreline of support or}$$

$$12\phi + d/2 \text{ from the face of the support}\} \qquad \text{Eq. (4.5a)}$$

where d is the effective depth of the beam.

It is interesting that a deep section can be assumed to be a simply supported beam in common design practices in Hong Kong, although such deep sections could induce considerable negative moments due to partial fixity. Engineers are reminded to design for negative moments, see Example 1 in a later section of this chapter.

Simplified rules for simply supported or the end support of continuous solid slabs (Cl. 9.3.1.3)

A general detailing rule recommended for simply supported solid slabs, as stipulated in HKBD 2013, is to resist 50% of the maximum mid-span moment and 50% of the calculated maximum span reinforcement for both the top and bottom reinforcements, respectively. The reinforcements are to be anchored into the support in accordance with Eqs. (4.2) and (4.5b).

If the design ultimate shear stress at the face of the support is less than half the appropriate value of the concrete shear stress capacity (v_c), Cl. 6.1.2.5 in HKBD 2013 recommends a straight length bar that extends beyond the centreline of the support that is equal to either one-third of the support width or 30 mm; whichever is greater can be considered as the effective anchorage.

The effective full tensile anchorage is assumed by providing the following simplified detailing rules:

$$l_{slab,sr} = \max\{0.15l_{span}, 45\phi\} \qquad \text{Eq. (4.5b)}$$

where l_{span} is the span length of the slab.

4.2.3 Design provisions for RA design procedures in EN 1992-1-1 (2004)

This section reviews the anchorage detailing provisions for cast-in straight longitudinal reinforcements in EN 1992-1-1 (2004), which uses the RA design procedure to design post-installed reinforcement applicable for system with a valid ETA in accordance with EAD 330087-00-0601.

Straight bar anchorage (Cl. 8.4)

An important difference in calculating the anchorage length in Cl. 8.4.3 of EN 1992-1-1 (2004) compared to the defunct BS 8110 Part 1 1997 (and HKBD 2013), is the consideration of design stress (f_{sd}) rather than simply assuming the maximum characteristic yield stress (f_{yk}) with a partial safety factor (γ_s) of steel. The assumption that a reinforcement can be fully stressed to its yield strength is rarely the case, as good detailing principles place lapped splices in areas of low stress, with the provided area being greater than the required area of the steel (The Concrete Centre, 2015). Although the design stress (f_{sd}) was not precisely described in the code, it can be rationally determined by using the ratio of the steel area required ($A_{s,rqd}$) to the steel area provided ($A_{s,prov}$), multiplied by the design yield strength of the steel (i.e., $A_{s,rqd}/A_{s,prov} \cdot f_{yk}/\gamma_s$) (The Concrete Centre, 2015). Structural designers are cautioned that a shorter anchorage length may cause other failure mechanisms associated with anchors, i.e., cone-shaped concrete breakout and bond failure (pull-out).

The design bond stress (f_{bd}) according to Cl. 8.4.2 (2), is a function of the design value of the concrete tensile strength (f_{ctd}) according to Cl. 3.1.6 (2)P. Eq. (4.6a) shows the bond stress estimation with coefficients η_1 and η_2 to implicitly account for the bond conditions, position of the reinforcement, and reinforcement diameter.

$$f_{bd} = 2.25\,\eta_1\eta_2 f_{ctd} \quad \text{(in EN 1992-1-1 (2004) format)} \qquad \text{Eq. (4.6a)}$$

Analogous to HKBD 2013, Eq. (4.6a) can be modified into Eq. (4.6b) for factored bond stress capacity (f_{bu}) with the inclusion of a material partial safety factor (γ_m).

$$f_{bd} = 2.25\,\eta_1\eta_2 f_{ctk,0.05}/\gamma_m \quad \text{(in HKBD 2013 format)} \qquad \text{Eq. (4.6b)}$$

where η_1 is a coefficient for the bond conditions and is related to the reinforcement position during concreting (1.0 for good and 0.7 for other); η_2 is a coefficient for the influence of the reinforcement diameter (1.0 for $\phi \leq 32$ mm and $(132 - \phi)/100$ for $\phi > 32$ mm); and f_{ctd} is equal to the characteristic tensile strength at 5% fractile ($f_{ctk,0.05}$) with consideration of a partial safety factor ($\gamma_m = 1.5$).

Hence, whilst the basic derivation in Eq. (4.1) is analogical, the stress experienced by the reinforcement is the design stress rather than the characteristic yield stress. Eq. (4.7) shows the basic required anchorage length ($l_{b,rqd}$).

$$l_{b,rqd} \geq \frac{f_{sd}}{f_{bd}}\frac{\phi}{4} \qquad \text{Eq. (4.7)}$$

Interestingly, EN 1992-1-1 (2004) introduced an additional procedure to check the design anchorage length (l_{bd}), and imposed a minimum anchorage length ($l_{b,min}$), both of which are not required in HKBD 2013. Eqs. (4.8) and (4.12) are the calculations for the design and minimum anchorage lengths, respectively.

$$l_{bd} = \alpha_1\alpha_2\alpha_3\alpha_5 l_{b,rqd} \geq l_{b,min} \qquad\qquad\text{Eq. (4.8)}$$

where α_1 is a coefficient for the effect of the form of the bars, assuming adequate cover (for straight bars, α_1 is 1.0 regardless of tension or compression action), and α_2 is a coefficient for the effect of the minimum concrete cover to consider splitting failure; shown in Eq. (4.9a) and (4.9b) for straight bars.

$$0.7 \leq \alpha_2 = 1 - \frac{0.15(c_d - \phi)}{\phi} \leq 1.0 \text{ (Tension)} \qquad\qquad\text{Eq. (4.9a)}$$

$$\alpha_2 = 1.0 \text{ (Compression)} \qquad\qquad\text{Eq. (4.9b)}$$

where $c_d = \min \{s/2, c_1, c_s\}$ for straight bars, s is the clear spacing of the bars, c_1 is the side cover, and c_s is the top or bottom cover.

Although the rest of the coefficients present challenges for post-installed reinforcement systems, they are still included here for discussion. Coefficient α_3 in Eq. (4.10) is used to account for the effects of confinement by transverse reinforcements, and coefficient α_5 in Eq. (4.11) is used for the effect of the pressure transverse to the plane of splitting along the design anchorage length.

$$0.7 \leq \alpha_3 = 1 - K\lambda \leq 1.0 \text{ (Tension)} \qquad\qquad\text{Eq. (4.10a)}$$

$$\alpha_3 = 1.0 \text{ (Compression)} \qquad\qquad\text{Eq. (4.10b)}$$

where K is defined in Figure 4.1 and λ is ratio of the excess transverse reinforcement area to the longitudinal reinforcement area, $(\Sigma A_{st} - \Sigma A_{st,min}) / A_s$.

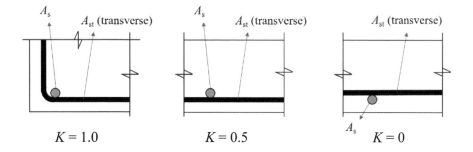

Figure 4.1: Values of K for beams and slabs in EN 1992-1-1 (2004)

$$0.7 \leq \alpha_5 = 1 - 0.04p \leq 1.0 \text{ (Tension)} \qquad\qquad\text{Eq. (4.11a)}$$

where p is the transverse pressure (in MPa) in the ultimate limit state along l_{bd}.

This review concerns only post-installed reinforcements; hence, unrelated coefficients are excluded, i.e., α_4 for welded transverse bars. The minimum anchorage length can be calculated by using Eqs. (4.12a) and (4.12b).

$l_{b,min} \geq max\{0.3l_{b,rqd}, 10\phi, 100 \text{ mm}\}$ (Tension)   Eq. (4.12a)

$l_{b,min} \geq max\{0.6l_{b,rqd}, 10\phi, 100 \text{ mm}\}$ (Compression)   Eq. (4.12b)

Note that the minimum anchorage length ($l_{b,min}$) shall be multiplied by an amplification factor (α_{lb}) to account for the difference between the cast-in place and post-installed reinforcements in cracked concrete. In general, if no tests are carried out for post-installed reinforcements in cracked concrete in accordance with EAD 330087-00-0601 (2018), α_{lb} is equal to 1.5.

Lapped splices (Cl. 8.7.3)

The design lap length in EN 1992-1-1 (2004) is calculated as follows:

$l_o = \alpha_1\alpha_2\alpha_3\alpha_5\alpha_6 l_{b,rqd} \geq l_{o,min}$ Eq. (4.13)

where α_1, α_2, α_3, and α_5 have already been defined (see Eqs. (4.9) to (4.11)). α_6 is a coefficient of the percentage of lapped bar (p_1) relative to the total cross-sectional area within 0.65 l_o from the centre of the lap length (see Eq. (4.14)).

$1.0 \leq \alpha_6 = (p_1/25)^{0.5} \leq 1.5$ Eq. (4.14)

The minimum lap length can be calculated by using:

$l_{o,min} \geq max\{0.3\alpha_6 l_{b,rqd}, 15\phi, 200 \text{ mm}\}$ Eq. (4.15)

Similar to the minimum anchorage length for post-installed reinforcements, the minimum lap length ($l_{o,min}$) shall be multiplied by an amplification factor (α_{lb}) to account for the difference between the cast-in place and post-installed reinforcements in cracked concrete. In general, if no tests are carried out for post-installed reinforcements in cracked concrete in accordance with EAD 330087-00-0601 (2018), α_{lb} is equal to 1.5.

Simplified rules for simply supported beams (Cl. 9.2.1.2(1) and Cl. 9.2.1.4(1))

Values of 15% of the maximum bending moment in the span and 25% (National Annex (of the Eurocode) dependent, versus 50% in HKBD 2013) of the steel area provided in the span are recommended for the top and bottom reinforcement, respectively, at the support of simply supported beams. Both the top and bottom steel are to be anchored with l_{bd}, measured from the face of the support. Note that Cl. 9.2.1.4(2) allows for the use of a strut-and-tie equivalent model to calculate the axial forces in the reinforcement, which appears to be more suitable for the design stress (f_{sd}) estimation in Eq. (4.7).

Simplified rules for simply supported solid slabs (Cl. 9.3.1.2)

In simply supported slabs, 15% (for end supports) to 25% (intermediate supports) of the maximum bending moment in the span and 50% of the calculated span reinforcement should be provided for the top and bottom bars for supporting solid slabs, respectively (as opposed to 50% in HKBD 2013). Both the top and bottom steel are anchored with l_{bd}, measured from the face of the support. As with simply supported beams, Cl. 9.2.1.4(2) allows for the use of STM.

4.2.4 Summary of RA design provisions

Table 4.3 shows an overview of a range of the bond strength values calculated in accordance with the two RA design procedures. The case is assumed for ribbed bars with $\phi \leq 32$, where the reinforcement is in a good position during concreting. The material partial safety factors for the bond stress (1.4 with HKBD 2013 and 1.5 with ETA) are excluded. In general, HKBD 2013 presents more conservative values (except for lower strength concrete with a cube strength of 20 MPa and $\alpha_2 = 1.0$) compared to EN 1992-1-1 (2004), as per ETA.

Table 4.3: Summary of bond strength in accordance with rebar anchorage design procedures in HKBD 2013 and EN 1992-1-1 (2004), as per ETA

Concrete characteristic cube strength, $f_{cu,k}$ (MPa)	Concrete characteristic tensile strength at 5% fractile, $f_{ctk,0.05}$ (MPa)	Bond strength (tension)				Bond strength (compression)	
		HKBD 2013 ($\beta = 0.5$)	ETA (normalised by $\alpha_2 = 0.7$*)	ETA (normalised by $\alpha_2 = 0.85$*)	ETA (normalised by $\alpha_2 = 1.0$*)	HKBD 2013 ($\beta = 0.63$)	ETA (normalised by $\alpha_2 = 1.0$*)
25	1.5	3.5	4.8	4.0	3.4	4.4	3.4
30	1.8	3.8	5.8	4.8	4.1	4.8	4.1
40	2.1	4.4	6.7	5.5	4.7	5.6	4.7
50	2.5	4.9	8.0	6.6	5.6	6.2	5.6
60	2.9	5.4	9.3	7.7	6.5	6.8	6.5

* α_2 is a coefficient for the effect of minimum concrete cover to consider splitting failure, and is stated in Eqs. (9a) and (9b) for straight bars. In this example, the limit boundary is equal to 0.7 and 1.0, respectively, where the case of $\alpha_2 = 1.0$ is more susceptible to splitting failure due to insufficient edge cover. Note that $\alpha_2 = 1.0$ corresponds to a concrete cover of 1 ϕ, which present challenges for hole drilling. The minimum concrete cover to account for possible deviations in drilling is provided in Table 4.5, with a minimum concrete cover of 2 ϕ, which corresponds to $\alpha_2 = 0.85$.

Note: Excludes material safety factor for bond stress, i.e., 1.4 for HKBD 2013 and 1.5 for EN 1992-1-1, 2014

4.3   A state-of-the-art moment connection design method for post-installed reinforcements using a strut-and-tie model

For cast-in-situ moment connections, reinforcements with a hook are usually terminated inside the connection. However, for moment connections that are constructed with post-installed reinforcements, it is impracticable to fix post-installed reinforcements with a hook into existing structures. Straight post-installed reinforcements must be used instead. Thus, post-installed bars rely on adhesive bonding rather than bearing at the hook to anchor the bars. If the anchorage length of the bar is insufficient, cone-shaped breakout failure may subsequently develop. A new design method for moment connections with post-installed reinforcements based on the STM was proposed in Kupfer et al. (2003) and validated by Bilal et al. (2006). The STM complies with the RC design procedures and DIN 1045-1 (2008) in that the internal tensile forces cannot be directly transferred to concrete if the design tensile strength (i.e., $f_{ctd} = \alpha_{ct} f_{ctk,0.05}/\gamma_m$) stipulated in EN 1992-1-1 (2004) is exceeded. Note that the coefficient α_{ct} is defined in the National Annex of the Eurocode and is usually equal to 1.

Figure 4.2 illustrates the STM for installing a new slab (or beam) onto an existing wall (or column) through a moment connection with post-installed reinforcements. The connection area can be defined as four zones. Zone 1 is the newly cast slab, whilst Zones 0, 2, and 3 are in the existing wall. The anchor of the post-installed reinforcement is placed in Zone 0, which is in between Zones 2 and 3.

As shown in Figure 4.2, the compressive strut exerts force F_o onto the anchor of the post-installed reinforcement. An effective strut-and-tie system begins with an anchorage length that is sufficiently long (for instance, greater than 15 ϕ). Normally, cone-shaped concrete breakout is designed at a prism angle of 35° according to BA design procedures, such as EN 1992-4 (2018). However, in moment connection cases, the formation of cone cracks will be inhibited by the compressive strut. Instead of the propagation of cone cracks, compressive strut failure or concrete splitting failure associated with the strut-and-tie system may occur. On the contrary, if the anchorage length is too short, the compressive stress field from the strut F_o cannot effectively prevent internal concrete cracking. Thus, cone shaped breakout failure may happen. In practice, however, reinforcement pull-out failure caused by bond failure is unlikely to occur, as high bond strength adhesives are often used in post-installed reinforcement systems.

The steps of the design procedure are as follows:

a. Calculate the force equilibrium at the joint
b. Determine the anchorage and embedment depths
c. Check the tension in the existing reinforcements
d. Check the concrete compressive strut
e. Check the splitting force in the transition zone

The anchorage length can be determined based on the adhesive bond strength. To formulate the STM, first, a strut angle θ that corresponds to $\theta = \tan^{-1}(z_0/z_{1r})$ (as shown in Figure 4.2 but limited to Eq. (4.20)) is selected, then the internal forces in the STM can be obtained by equilibrium. After this, the capacity of the STM components are to be checked.

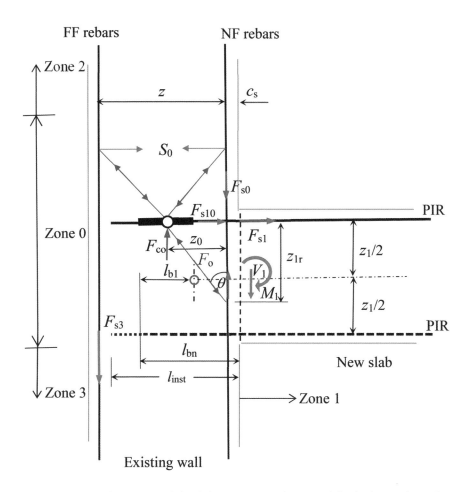

Figure 4.2: STM of moment resisting joint (note: FF and NF stand for far face and near face, respectively, whilst PIR stands for post-installed reinforcement)

Anchorage check

The compressive strut is anchored in the bond region of post-installed reinforcements. The tensile force in the post-installed reinforcements F_{s1} can be obtained from the applied moment M_1:

$$F_{s1} = M_1/z_{1r} \qquad\qquad\qquad \text{Eq. (4.16)}$$

where z_{1r} is the effective inner lever arm which can be determined from EN 1992-1-1 (2004), as shown in Eq. (4.17a) (Narayanan and Goodchild, 2006).

$$z_{1r} = d\,[1 + (1 - 3.529\,K)^{0.5}]/2 \quad \text{and} \quad K = M_1/(f_{ck} \cdot b \cdot d^2) \qquad \text{Eq. (4.17a)}$$

where d is the effective depth, b is the structural width, and f_{ck} is the characteristic compressive cylinder strength of concrete at 28 days.

Similarly, the effective inner lever arm can also be obtained from HKBD 2013, as shown in Eq. (4.17b).

$$z_{1r} = d\,[0.5 + (0.25 - K / 0.9\,)^{0.5}] \quad \text{and} \quad K=M_1/(f_{cu}{\cdot}b{\cdot}d^2) \qquad \text{Eq. (4.17b)}$$

Furthermore, if maximum F_{s1} is needed, the minimum z_{1r} in opening and closing moment cases may be taken as $0.85d$ and z_1, respectively (Kupfer et al., 2003).

For $F_{s10} = F_{s1}$, along with the sum of the reinforcement perimeters ($\sum u$), the effective anchorage length (l_{b1}) is:

$$l_{b1} \geq \frac{F_{s10}}{f_b {\cdot} \sum u} \qquad \text{Eq. (4.18)}$$

where f_b is the bond strength of the adhesion system, qualified as per EAD 330087-00-0601 (2018).

In the joint region (Zone 0), the lever arm of the internal forces z_0 is obtained by subtracting the concrete cover (c_s) and half of the design anchorage length (l_b) from the design anchorage depth (l_{bn}):

$$z_0 = l_{bn} - c_s - \phi/2 - l_{b1}\,/2 \geq l_{b1}\,/2 \qquad \text{Eq. (4.19a)}$$

However, if the actual installed anchorage depth (l_{inst}) is much deeper than required, a more realistic z_0 shall be calculated from the strut angle θ,

$$z_0 = z_{1r}\,\tan\theta \qquad \text{Eq. (4.19b)}$$

Noting that the strut angle θ is usually limited to

$$30° < \theta \leq 60° \qquad \text{Eq. (4.20)}$$

In such a case, the design anchorage depth can be calculated as:

$$l_{bn} = c_s + \phi/2 + z_0 + l_{b1}\,/2 \qquad \text{Eq. (4.21)}$$

Reinforcement check for near face of wall

When considering the moment equilibrium of the vertical forces in a section of wall, the reinforcement tensile force (F_{s0}) on the near face (NF) can be expressed as:

$$F_{s0} = M_1\,(1/z_0 - 1/z) \qquad \text{Eq. (4.22)}$$

where z is distance between the NF reinforcement and the far face (FF) reinforcement of the wall. The reinforcement area required is $A_{s0,rqd} = F_{s0}/(f_{yk}\,/g_s)$.

Reinforcement check for far face of wall

Similarly, the FF reinforcement tensile force F_{s3} is:

$$F_{s3} = M_1/z \qquad \text{Eq. (4.23)}$$

Given the area of the FF reinforcement (A_{s3}), the tensile stress in the reinforcement can be obtained ($f_{s3} = F_{s3}/A_{s3}$). In fact, this calculation is the same as that for the bending check in conventional RC design procedures.

Compressive strut check

As shown in Figure 4.2, the vertical force component (F_{c0}) of the strut force (F_o) in the joint region is acting on the centre of the required anchorage length ($l_{b,rqd}$) and can be expressed as,

$$F_{c0} = \frac{M_1}{z_0}$$ Eq. (4.24)

According to Section 6.5 in EN 1992-1-1 (2004), the maximum strength of a concrete strut is calculated by using the splitting reduction factor (α_{cc}) for the reduced concrete strength in the joint region:

$$f_R = \alpha_{cc} f_{cd}$$ Eq. (4.25a)

$$\alpha_{cc} = k_2 \, v' = k_2 \, (1 - f_{ck}/250)$$ Eq. (4.25b)

where k_2 is taken as 0.85 in compression-tension nodes. Finally, the maximum strut resistance (F_R) must be higher than the internal strut force, i.e.,

$$F_R \geq F_o.$$ Eq. (4.26)

$$F_o = F_{c0}/\cos \theta$$ Eq. (4.27)

$$F_R = \alpha_{cc} f_{ck}/Y_c \, (b \cdot l_{b1} \cdot \cos \theta)$$ Eq. (4.28)

where Y_c is the material partial safety factor for concrete and $l_{b1} \cdot \cos \theta$ is the width of the strut. Alternatively, the linear strut efficiency factor v' can be replaced by using a hyperbolic equation as proposed in the fib Model Code for Concrete Structures (fédération internationale du béton, 2013); i.e., $\alpha_{cc} = (0.75\eta_{fc})$ with $\eta_{fc} = (30/f_{ck})^{1/3} \leq 1$.

Splitting tensile stress in the discontinuity zone

In the wall region of the STM, the horizontal component (S_o) of the struts contributes to the tensile stress within the concrete that causes splitting failure. The maximum splitting moment (M_{sp}) is determined from the transverse bursting stresses in the anchorage zone. After obtaining the section modulus (W_{sp}), the splitting stress can be calculated with $f_{sp} = M_{sp}/W_{sp}$, based on Kupfer et al. (2003).

$$f_{sp} = F_{c0} \cdot z_0 \cdot \left(1 - \frac{z_0}{z}\right) \cdot \left(1 - \frac{l_{b1}}{2z}\right) / \frac{b \cdot z^2}{2.42}$$ Eq. (4.29a)

f_{sp} is to be checked against the concrete characteristic tensile strength at 5% fractile ($f_{ctk,0.05}$), which can be experimentally determined or obtained indirectly from the concrete compression strength (f_{ck}) (Cl. 3.1.6(2) of EN 1992-1-1).

$$f_{sp} \leq f_{ctk,0.05}$$ Eq. (4.29b)

4.3.1 Confinement method to increase bond strength (extension of EN 1992-1-1 (2004))

When the confinement provided by concrete is high, for instance, when the minimum concrete cover (c_d) is larger than ($3\,\phi$), it is possible to use the full adhesive bond strength (Randl and Kunz, 2012) instead of the bond strength given in EN 1992-1-1 (2004). Pull-out failure rather than splitting failure will be considered in the design. The bond strength of post-installed and cast-in reinforcements will no longer be similar. In fact, the former will be much higher than the latter. When this is the case, the splitting failure factor (α_2) in Eq. (4.9a) can be replaced by a pull-out failure factor (α_2') (see Eq. (4.30a)) in order to extend the linear relation of the design bond strength from f_{bd} to $f_{bd,sp}$ above the design limit given in EN 1992-1-1 (2004) (see Eq. (4.30b) and Figure 4.3). As a result, the anchorage length due to the splitting bond strength ($l_{bd,sp}$) for $c_d/\phi > 3$ will be shorter. However, more research is required to further elaborate on the testing method in order to determine a design bond strength coefficient δ for different adhesive products since the validity of the proposal by Kunz and Randl (2012) is limited to the tested product and conditions. Such a coefficient accounts for different cover thicknesses and adhesive types.

$$\alpha_2' = \frac{1}{\frac{1}{0.7}+\delta\cdot\frac{c_d-3\phi}{\phi}} \geq 0.25 \qquad\qquad \text{Eq. (4.30b)}$$

$$f_{bd,sp} = f_{bd}/\alpha_2' \leq f_{bd,p} \qquad\qquad \text{Eq. (4.30a)}$$

where $f_{bd,p}$ is the characteristic bond strength associated with pull-out failure that can be determined from a relevant anchoring injection system with product performance approved by ETA.

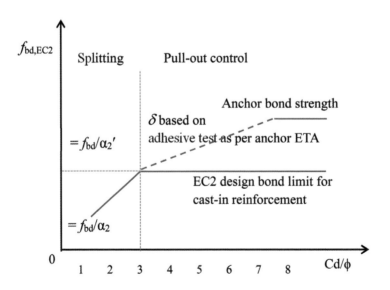

Figure 4.3: Effect of concrete cover on bond strength

4.3.2 Design provisions in HKBD (2013) to supplement STM

This section reviews the anchorage detailing provisions for cast-in straight longitudinal reinforcements in HKBD 2013, which supplements the use of STM for the design of post-installed reinforcements. Similar to the RA design procedures in HKBD 2013, equations that govern the RA design can be applied to the STM. In addition to all forces being factored, material partial safety factors shall also be incorporated. The factored bond stress capacity (f_{bu}) obtained from the concrete characteristic cube strength given in Eq. (4.3) is used for the adhesive bond strength (f_b).

Straight bar anchorage

When determining the effective anchorage length (l_{b1}) from the design tensile force of the post-installed reinforcement ($F_{s10,d}$), Eq. (4.18) is modified as follows:

$$l_{b1} = \frac{F_{s10,d}}{f_{bu} \cdot \Sigma u} \qquad\qquad \text{Eq. (4.31)}$$

where Σu is the sum of the reinforcement perimeters. This anchorage bond length must be shorter than the design anchorage bond length (l_b) obtained from the RA design procedures with Eq. (4.2), and the strut angle should not exceed the limit as calculated by Eq. (4.20).

The reinforcement should be checked with the design yield strength (incorporated with a material safety factor, $\gamma_s = 1.15$) obtained with Eqs. (4.22) and (4.23). For the concrete strut resistance (F_R) determined with Eq. (4.28), a concrete partial safety factor (γ_c) of 1.5 shall be used. The parameters for α_{cc} and k_2 are equal to 1.0 and 0.85, respectively, as per Cl. 3.1.6(1) and Cl. 6.5.2(2) of EN 1992-1-1 (2004). The concrete characteristic tensile strength ($f_{ctk,0.05}$) values to check concrete splitting with Eq. (4.29b) are given in Table 4.3.

Lapped splices

Eq. (4.4) and the relevant factors for tension and compression cases are applicable to determine the minimum lapped splice length (l_o).

4.3.3 Design provisions in EN 1992-1-1 (2004) to supplement STM

For STM designs, calculating the required anchorage bond length ($l_{b,rqd}$) can be done based on the RA design procedures in Cl. 8.4.4 of EN 1992-1-1 (2004). Hence, the design bond stress (f_{bd}) and the factored bond stress capacity (f_{bu}) as calculated by Eq. (4.6) are used to determine the design anchorage length (l_{bd}). The design anchorage length (l_{bd}) can be derived from the required anchorage length ($l_{b,rqd}$) but is subject to the minimum anchorage length ($l_{b,min}$) (Eqs. (4.7) to (4.12)).

If the existing structure is not deep enough, the maximum possible installed length can be obtained by following EAD 330232-00-0601 (2016), which is intended for use of mechanical fasteners in concrete and can be determined by using:

$$l_{inst,max} \geq h - \max\{2d_o, 30 \text{ mm}\} \qquad\qquad \text{Eq. (4.32)}$$

where h is the depth of the section and d_o is the drilled hole diameter.

If the installed anchorage depth (l_{inst}) is much longer than that required by the STM, it is more rational to use the effective anchorage depth (l_{bn1}) rather than the embedded length up to the bar end (i.e., Eq. (4.32)). A simpler approach can be taken by assuming that the strut angle $\theta = 60°$. Then, the effective anchorage length can be determined by using Eq. (4.33), where all of the symbols are defined in Figure 4.2.

$$l_{bn} = c_s + z_{1r} \tan 60° + l_{b1} / 2 \qquad\qquad \text{Eq. (4.33)}$$

Both the checking process of the reinforcement and compressive strut remains the same as stated in the previous section. With the concrete characteristic tensile strength ($f_{ctk,0.05}$) obtained from EN 1992-1-1 (2004), the splitting tensile stress can be checked by using:

$$f_{sp} \leq \alpha_{ct} f_{ctk,0.05} / \gamma_c \qquad\qquad \text{Eq. (4.34)}$$

where α_{ct} is equal to 1.0 based on Cl. 3.1.6(2) of EN 1992-1-1 (2004), the concrete material safety factor (γ_c) is 1.5, and $f_{ctk,0.05} = 0.7 f_{ctm} = 0.7 (0.3 f_{ck}^{2/3})$.

4.4   Recommended post-installed reinforcement design for Hong Kong

The RA and BA design procedures were individually developed based on different principles. In view of the general acceptance and familiarity of engineers in Hong Kong who use the RA design procedures for cast-in-situ concrete, modified design procedures for post-installed reinforcements will be introduced here. However, if readers are interested in the BA design procedures, a comparison of them in Chapter 17 of ACI 318 (2014) and CEN/TS 1992-4-5 (2009) (a document that was relevant prior to EN 1992-4 (2018)) can be found in Genesio et al. (2017b). A comprehensive review of the post-installed reinforcement designs which use the RA design procedures in EN 1992-1-1 (2004) and BA design procedures in CEN/TS 1992-4-5 (2009) is available in Mahrenholtz et al. (2015).

4.4.1 Design parameters

Following the RA design procedures in HKBD 2013, a simple and typical case is presented below to demonstrate the current challenges in designing post-installed reinforcement systems on walls using the current procedures.

Assuming $f_{cu,k} = 45$ MPa, $f_{yk} = 500$ MPa, β is 0.50 for tension stress, and the bar diameter ϕ varies from 12 mm to 32 mm:

Eq. (4.3) yields: $f_{bu} = \beta\sqrt{f_{cu}} = 0.5\sqrt{45} = 3.35$ MPa

Substitute this equation into Eq. (4.2). The calculated results are provided in Table 4.4.

The results in Table 4.4 show that in the case of slab-wall connections, a substantially long anchorage length is often necessary based on the assumption that the reinforcement is stressed to its yield strength. This assumption was revised in EN

Table 4.4: Calculating required anchorage length based on HKBD 2013

ϕ (mm)	l_b (mm) in Eq. (4.2) to nearest 5 mm	Remarks on beam-column connection	Remarks on slab-wall connection
12	390	Constructible, provided the column sectional depth is sufficient	More critical than beam-column connections, due to the limited thickness of the wall
16	520		
20	650		
25	815		
32	1040		

1992-1-1 (2004) (see Eq. (4.7)), in which the yield strength has been replaced by the design stress.

Six proposals are offered in this Guide to circumvent the issues of long anchorage length without compromising the safety of the connection.

Proposal 1: Provide a detailed option to determine the bond stress capacity (f_{bu}) of adhesives in post-installed reinforcement systems

A comparison of the bond stress (without a material safety factor) that is calculated by using HKBD 2013 versus EN 1992-1-1 (2004), as per ETA in Table 4.3, shows that HKBD 2013 generally provides a more conservative bond strength. Engineers who are already familiar with HKBD 2013 and wish to err on the side of caution can calculate the bond stress with HKBD 2013. Nevertheless, the use of EN 1992-1-1 (2004) is recommended here for engineers who wish to have more flexibility in design. This more detailed method improves the bond stress performance, even when splitting failure with a conservative factor $\alpha_2 = 1.0$ is taken into consideration. The calculation of f_{bu} is recommended as:

$$f_{bu} = \beta\sqrt{f_{cu}} \qquad\qquad \text{(general method)} \quad \text{Eq. (4.35a)}$$

$$f_{bu} = \text{ETA or manufacturer's technical data} \qquad \text{(detailed method)} \quad \text{Eq. (4.35b)}$$

Proposal 2: Replace the design yield stress (f_{yd}) with an actual design stress (f_{sd}) from a simple strut-and-tie equilibrium model.

The design stress (f_{sd}) is not precisely described in the design code in EN 1992-1-1 (2004). Although the Mineral Products Association (MPA) of the Concrete Centre published a statement (The Concrete Centre, 2015) saying that f_{sd} can be rationally determined by using the ratio of the steel area required ($A_{s,rqd}$) to the steel area provided ($A_{s,prov}$), multiplied by the design yield strength of the steel (i.e., $A_{s,rqd}/A_{s,prov} \cdot f_{yk}/\gamma_s$), it is in the opinion of the Guide that this definition still relies on the yield strength. Interestingly, Cl. 9.2.1.4(2) of EN 1992-1-1 (2004) allows for the use of the STM to calculate the axial forces (F_{Ed}) in the reinforcement (see Eq. (4.36)), which is well suited for estimating the design stress (f_{sd}).

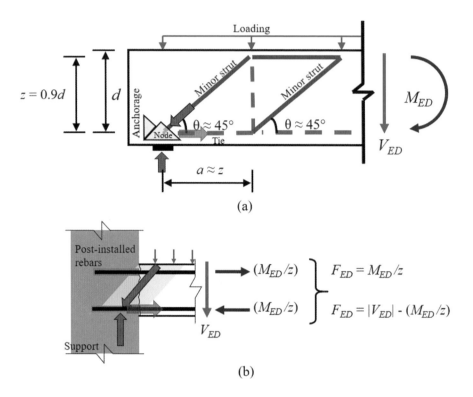

Figure 4.4: Idealised strut-and-tie model (a) in a simply supported beam, and (b) in a general case of post-installed rebar connection

$$F_{Ed} = |V_{Ed}|\frac{a}{z} + N_{Ed} \qquad\qquad \text{Eq. (4.36)}$$

where V_{Ed} is the design shear force, a is the shear span, z is assumed to be 0.9 d, d is the effective depth of the section, and N_{Ed} is the axial force (direct axial or resultant of bending) to be added to or subtracted from the tensile force. Figure 4.4 shows the idealised STM and simplification of Eq. (4.36), which assumes a 45-degree truss model, hence a/z is equal to unity.

The actual design stress (f_{sd}) can be estimated with:

$$f_{sd} = F_{Ed}/A_s = [|V_{Ed}| \pm M_{Ed}/z] / A_s \qquad\qquad \text{Eq. (4.37)}$$

where V_{Ed} is translated into tension force at the bottom bar at the support and zero tension at the top bar at the support, M_{Ed} depends on the assumption of the design of whether it is a simply supported or fixed support situation, z is assumed to be 0.9 d, and A_s is the reinforcement area at the top or bottom of the member.

Note that zero tension at the top bar for a simply supported member is an idealised assumption, and therefore should be reviewed based on the provided top bar, as per the minimum reinforcement percentage (i.e., 0.13% A_c) and detailing practices for partial

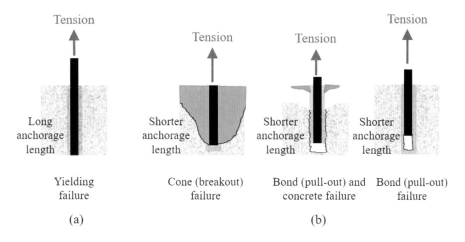

Figure 4.5: Failure modes: (a) anchorage length designed to yield strength (f_{yk}), and (b) shorter anchorage length designed to actual design strength (σ_{sd})

fixity (i.e., 15% of the mid-span moment for beam, as discussed in Section 4.2.2 for the RA design procedures in HKBD 2013). Note that HKBD 2013 recommends 50% of the mid-span moment should be provided as the top bar for simply supported slabs. Nevertheless, the detailing for partial fixity stated in Cl. 9.3.1.2(2) of EN 1992-1-1 (2004) recommends that end support moments to be resisted may be reduced to 15% of the maximum moment in the adjacent span for slabs to be resisted by the top bar.

Proposal 3: Impose a minimum anchorage length ($l_{b,min}$)

Proposals 2 and 3 should be applied together. However, readers should be aware that although allowing a design stress (f_{sd}) to be lower than the yield stress of the steel (f_{yk}) may give the advantage of a shorter anchorage length, it will also trigger other failure modes associated with the BA design procedures, such as cone-shaped concrete breakout (note that cone failure can be predicted if the tensile strength of concrete is assumed in place) and bond failure (pull-out) (see Figure 4.5).

Hence, it is sensible to impose a minimum anchorage length ($l_{b,min}$) to prevent the initiation of these failure modes. Note that there are no provisions for minimum anchorage length, only a minimum lap length ($l_{o,min}$) is stipulated in Cl. 8.7.3 of HKBD 2013. Eq. (4.38) is proposed based on the minimum length requirement with reference to EN 1992-1-1 (2004) (i.e., Eq. (4.12)).

$$l_{b,min} \geq \alpha_{lb} \times \max\{0.3l_{b,rqd}, 10\phi, 100\text{ mm}\} \text{ (Tension)} \qquad \text{Eq. (4.38a)}$$

$$l_{b,min} \geq \alpha_{lb} \times \max\{0.6l_{b,rqd}, 10\phi, 100\text{ mm}\} \text{ (Compression)} \qquad \text{Eq. (4.38b)}$$

where $l_{b,min}$ is measured from the face of the support, the amplification factor for the minimum anchorage length (α_{lb}) is equal to 1.5 if no testing is carried out on post-installed reinforcements in cracked concrete in accordance with EAD 330087-00-0601

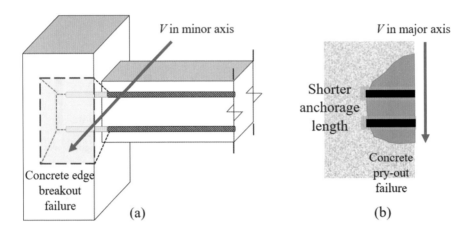

Figure 4.6: Precluded shear failure: (a) concrete edge failure, and (b) concrete pry-out failure

(2018). Similarly, the amplification factor should be considered for the minimum lap length ($l_{o,min}$) in Eq. (4.15).

If $l_{b,min}$ does not provide the distance from the centreline of the supporting members (i.e., columns or shear walls), the supporting members should be checked for additional moments caused by the eccentricity on the support.

Proposal 4: Check shear-induced web crushing

Provided that the shear (V in the major axis) is acting in-plane along the depth of the connecting member or along the length of the supporting member, and not out-of-plane along the edge (V in the minor axis), concrete edge failure will not occur. The minimum anchorage length ($l_{b,min}$) which inhibits the possibility of cone failure will be effective for preventing concrete pry-out failure (see Figure 4.6).

However, it is recommended that a check is done for web crushing by using one of the two methods illustrated in Figure 4.7; namely, (Method A) enhancing shear strength near supports as stipulated in Cl. 6.1.2.5(g) of HKBD 2013 or (Method B) checking the strut band stress in the strut-and-tie method.

Method (1): Enhancing concrete shear stress capacity near supports as provided in Table 6.3 and Cl. 6.1.2.5(g) of HKBD 2013:

$$v_{c,enhanced} = \left(\frac{2d}{a}\right) v_c = 2(0.79) \left(\frac{100A_s}{b_v d}\right)^{\frac{1}{3}} \left(\frac{400}{d}\right)^{\frac{1}{4}} \left(\frac{1}{\gamma_m}\right) \left(\frac{f_{cu}}{25}\right)^{\frac{1}{3}} \leq v_{applied} \quad \text{Eq. (4.39)}$$

where a 45-degree truss model is assumed; hence $d = a$.

Method (2): Checking struts by using the STM:

Assuming that the outer strut width reaches the intersection of the top of the connecting member with the face of the supporting member via a 45-degree angle, the strut band width (w_{strut}) can be estimated by using:

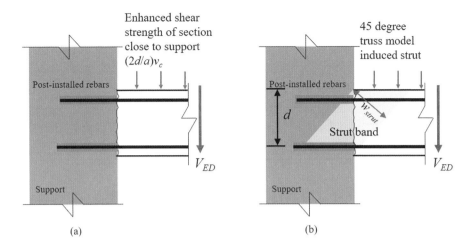

Figure 4.7: Check for shear-induced web crushing: (a) enhanced shear strength near support check, as per HKBD 2013, and (b) strut capacity check using strut-and-tie model

$$w_{strut} = d/\sqrt{2}$$ Eq. (4.40)

Hence the strut capacity is calculated with:

$$f_{strut} = \frac{V_{Ed}}{b\,w_{strut}} \leq \beta_{strut}\,f_{cu,k}/\gamma_c$$ Eq. (4.41)

where b is the width of the connecting member, β_{strut} is the recommended strut efficiency factor equal to $\beta_{strut} = 0.5$ for cube strength or 0.6 for cylinder strength according to Su and Looi (2016), and γ_c which is the material partial safety factor for concrete = 1.5.

Proposal 5: Impose a minimum edge distance

When reinforcements are placed at the edges, radial cracks may propagate through the cover (see Figure 4.8 on p. 52).

With reference to Eq. (4.9) in EN 1992-1-1 (2004), an α_2 factor is recommended to account for the effect of the minimum concrete cover by considering splitting failure. When α_2 is equal to 0.7, an explicit expression can be derived from α_2:

$$c_d \geq 3\phi$$ Eq. (4.42)

where $c_d = \min\{s/2, c_1, c_s\}$ for straight bars, s is the clear spacing of the bars, c_1 is the side cover, and c_s is the top or bottom cover.

EN 1992-1-1 (2004) states that the maximum boundary is reached when α_2 is equal to 1.0, and c_d corresponds to 1 ϕ. Note that such a small cover of 1 ϕ may present challenges with drilling holes in post-installed reinforcement systems. Hence, EAD 330087-00-0601 (2018) proposes a minimum cover as a function of the drilling method, reinforcement size, and with or without the use of a drilling aid, to take into account the possible deviations during the drilling process. Table 4.5 presents the relevant tables from EAD 330087-00-0601 (2018).

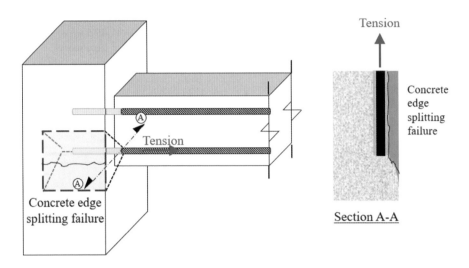

Figure 4.8: Splitting failure

Table 4.5: Minimum concrete cover (c_{min}) proposed in EAD 330087-00-0601 (2018)

Use of drilling aid	Drilling method	Bar diameter ϕ	c_{min}
No	Hammer or diamond	< 25 mm	30 mm + 0.06 $l_v \geq 2\phi$
		$\geq$ 25 mm	40 mm + 0.06 $l_v \geq 2\phi$
	Compressed air	< 25 mm	50 mm + 0.08 l_v
		$\geq$ 25 mm	60 mm + 0.08 $l_v \geq 2\phi$
Yes	Hammer or diamond	< 25 mm	30 mm + 0.02 $l_v \geq 2\phi$
		$\geq$ 25 mm	40 mm + 0.02 $l_v \geq 2\phi$
	Compressed air	< 25 mm	50 mm + 0.02 l_v
		$\geq$ 25 mm	60 mm + 0.02 $l_v \geq 2\phi$

Note: l_v is the setting anchorage depth of rebars (in mm).

Proposal 6: Use the STM for moment connections

In order for the post-installed reinforcements to allow moment connections, the STM design procedures are used to check the existing structures. Moreover, the effective anchorage length (l_{b1}) based on Eq. (4.33) needs to be longer than the minimum length ($l_{b,min}$) (see Eq. (4.38a)), as shown in the following:

$$l_{b1} \geq l_{b,min}$$ Eq. (4.43)

4.4.2 Design examples

The proposed design procedures will be illustrated through practical examples in this section. Four design examples will be discussed; namely, Example 1: A simply supported RC beam connected to an RC column, Example 2: A simply supported RC slab connected to an RC shear wall, Example 3: A moment connection of an extension of a new RC slab connected to an existing RC slab with lap splicing, and Example 4: A moment connection of an RC slab on a wall (based on the state-of-the-art design method that uses the STM).

Example 1: A simply supported RC beam connected to an RC column

In a retrofitting project for a building, the original beam (with pinned support detailed in accordance with HKBD 2013) installed in the internal bay of a frame is seriously degraded and damaged. A new RC beam is designed which would be connected to the existing RC column with post-installed reinforcements, as shown in Figure 4.9.

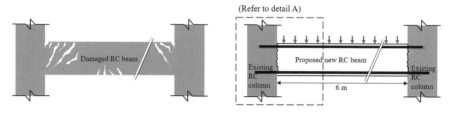

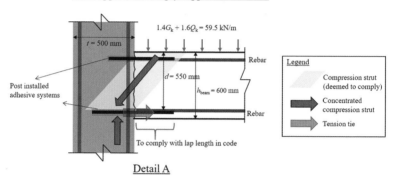

Detail A

Figure 4.9: Proposed post-installed rebar system for damaged simply supported RC beam

Structure dimensions, material, and load

Beam: $l_n = 6$ m, $h_{beam} = 600$ mm, $b = 300$ mm, cover $= 30$ mm, $d = 550$ mm, and $a_v = d$

Slab: $h_{slab} = 150$ mm, transverse span length $= 3$ m

Column: $h_{col} = 500$ mm, cover $= 35$ mm

Concrete strength grade: C40 (cube), $f_{ctk,0.05} \approx 2.1$ MPa

Reinforcement: $f_{yk} = 500$ N/mm^2, $\gamma_s = 1.15$

Dead loads (self-weight of beam and slab): $g_k = (24.5$ kN/m$^3 \times h_{beam} \times b) + (24.5 \times h_{beam} \times b) = (24.5 \times 0.6 \times 0.3) + (24.5 \times 0.15 \times 3) = 15.4$ kN/m

Superimposed dead loads (screeding, tiles, electrical, and partition walls): $g_k = (2.7$ kN/m$^2 \times 3) = 8.1$ kN/m

Live loads: $q_k = 5$ kN/m$^2 \times 3 = 15$ kN/m

Design load combination: At ultimate limit state (ULS), $S_d = (1.40\ g_k + 1.60\ q_k) = 56.9$ kN/m

(Cl. 2.3.2, HKBD 2013)

Structural analysis (design forces): At mid-span, $M_{Ed} = S_d \, l_n{}^2 / 8 = 256$ kNm

At support, $V_{Ed} = S_d \, l_n / 2 = 171$ kN

Predesigned beam

Bottom reinforcement required: At mid span, $A_{s,rqd,m} = M_{Ed} / (0.9d f_{yk}/\gamma_s) = 1191$ mm²

Reinforcement provided: At mid span, 3 nos. $\phi25$, $s = 90$ mm; $A_{s,prov,m} = 1473$ mm²

Deemed-to-comply check for concrete web crushing – Proposal 4:

Method 1: enhanced shear check

Concrete shear stress capacity,

$$v_c = 0.79 \left(\frac{100A_s}{b_v d}\right)^{\frac{1}{3}} \left(\frac{400}{d}\right)^{\frac{1}{4}} \left(\frac{1}{\gamma_m}\right) \left(\frac{f_{cu}}{25}\right)^{\frac{1}{3}} = 0.66 \text{ MPa}$$

(Table 6.3, HKBD 2013)

Assuming a 45-degree strut relative to the bottom longitudinal tensile reinforcement

$2d / a_v = 2$

Concrete enhanced shear stress capacity,

v_c enhancement at support $= 0.66(2) = 1.31$ MPa $<$ lesser of ($0.8\sqrt{f_{cu}} = 5.1$ MPa, 7 MPa)

(Cl. 6.1.2.5(g), HKBD 2013)

Applied shear stress,

$V_{Ed} = 171$ kN

$v_{Ed} = \frac{V_{Ed}}{b_v d} = 1.04$ MPa < 1.31 MPa, and thus acceptable.

Method 2: strut check using the STM

Assuming a 45-degree strut relative to the bottom longitudinal tensile reinforcement

$F_o = |V_{Ed}| / \sin 45° = 242$ kN

Assuming the outer strut width reaches the intersection of the top of the slab with the face of the supporting wall via a 45-degree angle, hence the strut width is

$w_{strut} = h/\sqrt{2} = 600/\sqrt{2} = 424$ mm.

$f_o = \frac{F_{strut}}{b w_{strut}} = 1.9$ MPa $< 0.50 f_{cu} / 1.5 = 13.3$ MPa, and thus acceptable.

(Su, R.K.L and Looi, D.T.W. (2016). "Revisiting the Unreinforced Strut Efficiency Factor", ACI Structural Journal, 113(2), pp. 301–312.)

Check for the anchorage length at the top and bottom post-installed reinforcements at the support, as shown in Figure 4.10.

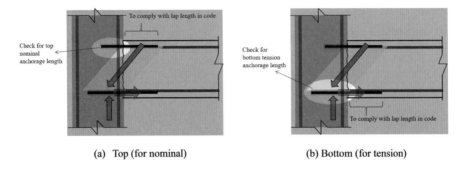

(a) Top (for nominal)   (b) Bottom (for tension)

Figure 4.10: Design check for anchorage length of post-installed reinforcement at RC column support

Top support	Bottom support
(1) Bond stress – Proposal 1	

The top and bottom supports are in tension.

General method based on Eq. (4.35a)

$f_{bu} = \beta\sqrt{f_{cu}} = 0.5\sqrt{40} = 3.2$ MPa

Note that the partial safety factor for bond stress = 1.4 is embedded.

Detailed method based on Eqs. (4.6), (4.9), and (4.35b)

Although the cover can be equal to 3ϕ for $\alpha_2 = 0.70$ to maximise the bond strength, it is, however, conservatively equal to 2ϕ, which implies that $\alpha_2 = 0.85$ to account for drilling deviations, as per the recommendation in EAD 330087-00-0601 (2018); reinforcement position is good during concreting and reinforcement diameter is less than 32 mm.

$$f_{bu} = \frac{f_{bd}}{\alpha_2} = 2.25\, \eta_1\eta_2 f_{ctd}/\alpha_2 = 2.25\, \eta_1\eta_2 f_{ctk,0.05}/\gamma_m/\alpha_2$$

$$= 2.25(1)(1)(2.1)/1.5/0.85 = 3.7 \text{ MPa}$$

| **(2) Tension tie force to be anchored – Proposal 2** | |

| Using Eq. (4.36), shear (V_{Ed}) will not transfer tension to the top bar and the end moment is zero due to the simply supported assumption: | Using Eq. (4.36), shear (V_{Ed}) will induce a direct tension via a strut of 45-degrees to the bottom bar. The end moment is zero due to the simply supported assumption: |

$f_{sd} = F_{Ed}/A_s = [|V_{Ed}| \pm M_{Ed}/z] / A_s$

$f_{sd} = F_{Ed}/A_s = [|V_{Ed}| \pm M_{Ed}/z] / A_s$

Nevertheless, the simplified rule is applied to the beams (see next row in Item 3).

Hence, $F_{Ed,tens} = |V_{Ed}| + 0 = 171$ kN

Post-installed steel area required:

$A_{s,rqd} = F_{Ed} / (f_{yk} / \gamma_s) = 393$ mm²

(3) Minimum post-installed reinforcements to be anchored at support – Proposal 2

Simplified rules (15% of the maximum bending in the span) for simply supported beam to control cracking at negative moments due to partial fixity:

$A_{s,\text{simplified rules}} = 0.15\, M_{Ed,\text{ mid-span}} / (0.9d\, f_{yk}/ \gamma_s) = 179$ mm²   (Cl. 9.2.1.5, HKBD 2013)

At the simply supported end of a member, half the calculated mid-span bottom reinforcement should be anchored:

$A_{s,\text{simplified rules}} = 0.50\, A_{s,\text{mid-span}} = 0.5(1191) = 595$ mm²   (Cl. 9.2.1.7, HKBD 2013)

Tension reinforcement, rectangular section subjected to flexure, and web in tension: For a rectangular slab with $f_{y,k} = 500$ MPa,

$A_{s,min} = 0.13\% \, A_c = 0.13/100\,(300 \times 600) = 234$ mm²   (Table 9.1, HKBD 2013)

(4) Decision for post-installed reinforcements to be anchored

$A_{s,req} = \max \{0, 179, 234\} = 234$ mm²

Reinforcement provided: 3 nos. ϕ12, $s = 90$ mm; and $A_{s,prov,m} = 339$ mm²

$A_{s,req} = \max \{393, 595, 234\} = 595$ mm²

Reinforcement provided: 3 nos. ϕ16, $s = 90$ mm; and $A_{s,prov,m} = 603$ mm²

(5) Anchorage length required

(Design for Proposal 2)

Back-calculate the equivalent hogging moment at the support:

$A_{s,\text{req top support}} / A_{s,\text{bottom mid-span}} \times M_{Ed} = 234/1191 \times 256 = 50$ kNm

Equivalent stress:

$f_{sd} = [50 / (0.9\ d)] \times 1000 / (3\ \pi\ 12^2/4) = 298$ MPa

General method

$l_{b,rqd} = \dfrac{f_{sd}}{f_{bu}} \dfrac{\phi}{4} = \dfrac{298}{3.2}(\dfrac{12}{4}) = 279$ mm

Detailed method

$l_{b,rqd} = \dfrac{f_{sd}}{f_{bu}} \dfrac{\phi}{4} = \dfrac{298}{3.7}(\dfrac{12}{4}) = 241$ mm

(If it is designed to HKBD 2013)

(Design for Proposal 2)

Using Eq. (4.7), $f_{sd} = 171000 / (3\,\pi\,16^2/4) = 283$ MPa

General method

$l_{b,rqd} = \dfrac{f_{sd}}{f_{bu}} \dfrac{\phi}{4} = \dfrac{283}{3.2}(\dfrac{16}{4}) = 358$ mm

Detailed method

$l_{b,rqd} = \dfrac{f_{sd}}{f_{bu}} \dfrac{\phi}{4} = \dfrac{283}{3.7}(\dfrac{16}{4}) = 306$ mm

(If it is designed to HKBD 2013).

From Eq. (4.1), $f_y = 500$ MPa

$$l_{b,rqd} = \frac{0.87 f_y \, \phi}{f_{bu}} \frac{\phi}{4} = \frac{0.87(500)}{3.2} \frac{12}{4}$$

$$= 413 \text{ mm}$$

Note that the anchorage length calculated as per HKBD 2013 almost reaches an RC column depth of 500 mm.

From Eq. (4.1), $f_y = 500$ MPa

$$l_{b,rqd} = \frac{0.87 f_y \, \phi}{f_{bu}} \frac{\phi}{4} = \frac{0.87(500)}{3.2} \frac{16}{4}$$

$$= 550 \text{ mm}$$

Note that the anchorage length calculated as per HKBD 2013 already penetrates an RC column depth of 500 mm.

(6) Minimum anchorage length – Proposal 3

Using Eq. (4.38a),

$l_{b,min} \geq \alpha_{lb} \times$

$\max\{0.3 l_{b,rqd}; 10\phi; 100 \text{ mm}\}$

General method

$l_{b,min} \geq 1.5 \times \max\{0.3(279) =$

$84; 10(12) = 120; 100\} \geq 180 \text{ mm}$

Detailed method

$l_{b,min} \geq 1.5 \times \max\{0.3(241)$

$$= 73; 10(12)$$

$$= 120; 100\} \geq 180 \text{ mm}$$

α_{lb} is equal to 1.5, assuming that no tests are carried out for post-installed reinforcements in cracked concrete in accordance with EAD 330087-00-0601 (2018).

Using Eq. (4.38a),

$l_{b,min}$

$\geq \alpha_{lb} \times \max\{0.3 l_{b,rqd}; 10\phi; 100 \text{ mm}\}$

General method

$l_{b,min} \geq 1.5 \times \max\{0.3(358) =$

$107; 10(16) = 160; 100\} \geq 240 \text{ mm}$

Detailed method

$l_{b,min} \geq 1.5 \times \max\{0.3(306)$

$$= 92; 10(16)$$

$$= 160; 100\} \geq 240 \text{ mm}$$

α_{lb} is equal to 1.5, assuming that no tests are carried out for post-installed reinforcements in cracked concrete in accordance with EAD 330087-00-0601 (2018).

(7) Provided anchorage length

General method

$l_b = \max \{l_{b,req}, l_{b,min}\} = \max\{279, 180\} = 279$ mm

Detailed method

$l_b = \max \{l_{b,req}, l_{b,min}\} = \max\{241, 180\} = 241$ mm

General method

$l_b = \max \{l_{b,req}, l_{b,min}\} = \max\{358, 240\} = 358$ mm

Detailed method

$l_b = \max \{l_{b,req}, l_{b,min}\} = \max\{306, 240\} = 306$ mm

(8) Minimum edge distance check for splitting failure (Proposal 5)

Based on Eq. (4.42),

$c_d = \min\{s/2, c_1, c_s\} = \min\{90/2, 35, 35\} = 35$ mm

Based on Eq. (4.42),

$c_d = \min\{s/2, c_1, c_s\} = \min\{90/2, 35, 35\} = 35$ mm

From Table 4.5* (with drilling aid, hammer drilled, $\phi = 12$)

30 mm $+ 0.02\, l_v \geq 2\phi$

*Or based on the ETA-approved technical data of the material.

General and detailed methods

$30 + 0.02\,(250) = 35$ mm $\geq$
$2\phi = 2(12) = 24$ mm

From Table 4.5* (with drilling aid, hammer drilled, $\phi = 16$)

30 mm $+ 0.02\, l_v \geq 2\phi$

*Or based on the ETA-approved technical data of the material.

General method

$30 + 0.02\,(358) = 37$ mm $\geq$
$2\phi = 2(16) = 32$ mm

Detailed method

$30 + 0.02\,(306) = 36$ mm $\geq$
$2\phi = 2(16) = 32$ mm

Provided cover

$c_d = 35$
$c_d / \phi = 35/12 = 2.9 > 2*$
(*assuming $\alpha_2 = 0.85$)

$c_d = 35$
$c_d / \phi = 35/16 = 2.2 > 2*$
(*assuming $\alpha_2 = 0.85$)

(10) Summary of design for post-installed reinforcements

Use 3 T12 @ 90 mm, $A_{s,prov} = 339$ mm² with $l_b = 280$ mm (general method) or $l_b = 250$ mm (detailed method), and $c_d = 35$ mm.

Since l_b extends beyond the centreline of the support ($500/2 = 250$ mm), one does NOT need to check for additional moments caused by the eccentricity on the support.

Use 3 T16 @ 90 mm, $A_{s,prov} = 603$ mm² with $l_b = 358$ mm (general method) or $l_b = 306$ mm (detailed method), and $c_d = 35$ mm.

Since l_b extends beyond the centreline of the support ($500/2 = 250$ mm), one does NOT need to check for additional moments caused by the eccentricity on the support.

Example 2: A simply supported RC slab connected to an RC shear wall

During construction, plans are made for the casting of the RC slab after constructing the RC shear wall. Since no starter bar is used, post-installed reinforcements are considered. Post-installed reinforcements for the new RC slab (see Figure 4.11) are to be designed.

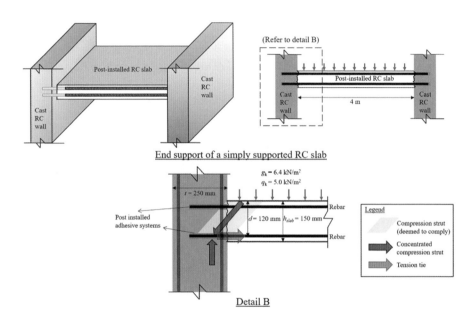

Figure 4.11: Details of end support using post-installed rebar system for simply supported RC slab

Structure dimensions, material, and load

Slab: $l_n = 4$ m, $h_{slab} = 150$ mm, $b = 1000$ mm (for per metre run), cover $= 30$ mm, $d = 120$ mm, and $a_v = d$

Shear wall: $h_{wall} = 250$ mm, cover $= 50$ mm, and $\phi 25$ vertical and horizontal bars with 250 mm spacings

Concrete strength grade: C35 (cube), $f_{ctk,0.05} \approx 1.95$ MPa

Reinforcement: $f_{yk} = 500$ N/mm^2, $\gamma_s = 1.15$

Dead loads (self-weight): $g_k = 24.5$ kN/m$^3 \times h = 24.5 \times 0.15 = 3.7$ kN/m^2

Superimposed dead loads (screeding, tiles, electrical, and partition walls): $g_k = 2.7$ kN/m^2

Live loads: $q_k - 5$ kN/m^2

Design load combination: At ULS, $S_d = (1.40\ g_k + 1.60\ q_k) = 16.9$ kN/m^2

(Cl. 2.3.2, HKBD 2013)

Structural analysis (design forces): At mid-span, $M_{Ed} = S_d \, l_n{}^2 \, / \, 8 = 33.9$ kNm/m

At support, $V_{Ed} = S_d \, l_n \, / \, 2 = 33.9$ kN/m

Predesigned slab

Bottom reinforcement required: At mid span, $A_{s,rqd,m} = $ MEd $ / \, (0.9d \, f_{yk}/\gamma_s) = 721$ mm²/m

Reinforcement provided: At mid span, $\phi 10$, $s = 100$ mm; $A_{s,prov,m} = 785$ mm²/m

Deemed-to-comply check for concrete web crushing – Proposal 4:

Method 1: enhanced shear check

Concrete shear stress capacity,

$$v_c = 0.79 \left(\frac{100 A_s}{b_v d}\right)^{\frac{1}{3}} \left(\frac{400}{d}\right)^{\frac{1}{4}} \left(\frac{1}{\gamma_m}\right) \left(\frac{f_{cu}}{25}\right)^{\frac{1}{3}} = 0.83 \text{ MPa}$$

(Table 6.3, HKBD 2013)

Assuming a 45-degree strut relative to the bottom longitudinal tensile reinforcement

$2d \, / \, a_v = 2$

Concrete enhanced shear stress capacity,

v_c enhancement at support $= 0.83(2) = 1.66$ MPa $<$ lesser of $(0.8\sqrt{f_{cu}} = 4.7$ MPa, 7 MPa)

(Cl. 6.1.2.5(g), HKBD 2013)

Applied shear stress,

$V_{Ed} = 33.9$ kN/m

$v_{Ed} = \frac{V_{Ed}}{b_v d} = 0.28$ MPa < 1.66 MPa, and thus acceptable.

Method 2: strut check using the STM

Assuming a 45-degree strut relative to the bottom longitudinal tensile reinforcement

$F_o = |V_{Ed}| \, / \sin 45° = 48$ kN/m

Assuming the outer strut width reaches the intersection of the top of the slab with the face of the supporting wall via a 45-degree angle, hence the strut width is

$w_{strut} = h/\sqrt{2} = 150/\sqrt{2} = 106$ mm.

$f_o = \frac{F_{strut}}{b w_{strut}} = 0.45$ MPa $< 0.50 \, f_{cu} \, / \, 1.5 = 11.7$ MPa, and thus acceptable.

(Su, R.K.L and Looi, D.T.W. (2016). "Revisiting the Unreinforced Strut Efficiency Factor", ACI Structural Journal, 113(2), pp. 301–312.)

Check for the anchorage length at the top and bottom post-installed reinforcements at the support, as shown in Figure 4.12.

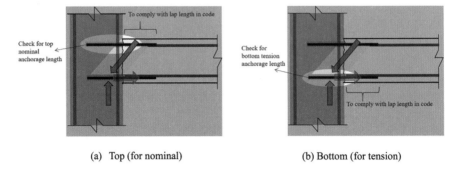

(a) Top (for nominal)   (b) Bottom (for tension)

Figure 4.12: Design check for anchorage length of post-installed rebar at RC wall support

Top support	Bottom support
(1) Bond stress – Proposal 1	

The top and bottom supports are in tension.

General method based on Eq. (4.35a)

$$f_{bu} = \beta\sqrt{f_{cu}} = 0.5\sqrt{35} = 2.96 \text{ MPa}$$

Note that the partial safety factor for bond stress = 1.4 is embedded.

Detailed method based on Eqs. (4.6), (4.9), and (4.35b)

Although the cover can be equal to 3ϕ for $\alpha_2 = 0.70$ to maximise the bond strength, it is, however, conservatively equal to 2ϕ, which implies $\alpha_2 = 0.85$ to account for drilling deviations, as per the recommendation in EAD 330087-00-0601 (2018); reinforcement position is good during concreting and reinforcement diameter is less than 32 mm.

$$f_{bu} = \frac{f_{bd}}{\alpha_2} = 2.25\, \eta_1\eta_2 f_{ctd}/\alpha_2 = 2.25\, \eta_1\eta_2 f_{ctk,0.05}/\gamma_m/\alpha_2$$

$$= 2.25(1)(1)(1.95)/1.5/0.85 = 3.44 \text{ MPa}$$

(2) Tension tie force to be anchored – Proposal 2	

| Using Eq. (4.36), shear (V_{Ed}) will not transfer tension to the top bar and the end moment is zero due to the simply supported assumption:

$f_{sd} = F_{Ed}/A_s = [\lvert V_{Ed}\rvert \pm M_{Ed}/z]\,/\,A_s$ | Using Eq. (4.36), shear (V_{Ed}) will cause direct tension via a strut of 45-degrees to the bottom bar. The end moment is zero due to the simply supported assumption:

$f_{sd} = F_{Ed}/A_s = [\lvert V_{Ed}\rvert \pm M_{Ed}/z]\,/\,A_s$ |

Nevertheless, the simplified rule is applied to the slabs (see next row in Item 3).

Hence, $F_{\text{Ed,tens}} = |V_{\text{Ed}}| + 0 = 33.9$ kN/m

Post-installed steel area required:

$A_{\text{s,rqd}} = F_{\text{Ed}} / (f_{\text{yk}} / \gamma_{\text{s}}) = 78$ mm²/m

(3) Minimum post-installed reinforcement to be anchored at support – Proposal 2

Simplified rules in EN 1992-1-1 (2004) (i.e., 15% of the maximum bending in the span) for the simply supported slab to control cracking at negative moments due to partial fixity:

$A_{\text{s,simplified rules}} = 0.15\, M_{\text{Ed, mid-span}} / (0.9d\, f_{\text{yk}}/ \gamma_{\text{s}}) = 108$ mm²/m

(Cl. 9.3.1.2(2), EN 1992-1-1 (2004))

Note that 50% of the maximum bending in the span for the simply supported slab in HKBD 2013 is not recommended here. The 50% assumption is similar to a fully fixed connection, somewhat contradictory to a simply supported slab, which may result in a longer anchorage length.

At the end support of the simply supported slab or continuous slab, half the calculated mid-span bottom reinforcement should be anchored:

$A_{\text{s,simplified rules}} = 0.50\, A_{\text{s,mid-span}} = 0.50\,(785) = 393$ mm²/m

(Cl. 9.3.1.3, HKBD 2013)

Tension reinforcement, rectangular section subjected to flexure, and web in tension: For a rectangular slab with $f_{\text{y,k}} = 500$ MPa,

$A_{\text{s,min}} = 0.13\%\, A_{\text{c}} = 0.13/100\, (1000 \times 150) = 195$ mm²/m   (Table 9.1, HKBD 2013)

(4) Decision for post-installed reinforcements to be anchored

$A_{\text{s,req}} = \max \{0, 108, 195\} = 195$ mm²

Reinforcement provided: 5 nos. $\phi10$, $s = 200$ mm; $A_{\text{s,prov,m}} = 393$ mm²/m

$A_{\text{s,req}} = \max \{78, 393, 195\} = 393$ mm²

Reinforcement provided: 5 nos. $\phi10$, $s = 200$ mm; $A_{\text{s,prov,m}} = 393$ mm²/m

(5) Anchorage length required

(Design for Proposal 2)

Back-calculate the equivalent hogging moment at the support:

$A_{\text{s,req top support}} / A_{\text{s,bottom mid-span}} \times M_{\text{Ed}} = 195/721 \times 33.9 = 9$ kNm/m

Equivalent stress:

$f_{\text{sd}} = [9 / (0.9\ d)] \times 1000 / (5\,\pi\,10^2/4) = 212$ MPa

(Design for Proposal 2)

From Eq. (4.7), $f_{\text{sd}} = 33900 / (5\,\pi\,10^2/4) = 86$ MPa

General method

$$l_{b,rqd} = \frac{f_{sd}}{f_{bu}}\frac{\phi}{4} = \frac{212}{2.96}(\frac{10}{4}) = 180 \text{ mm}$$

Detailed method

$$l_{b,rqd} = \frac{f_{sd}}{f_{bu}}\frac{\phi}{4} = \frac{212}{3.44}(\frac{10}{4}) = 154 \text{ mm}$$

(If it is designed to HKBD 2013)

Using Eq. (4.1), $f_y = 500$ MPa

$$l_{b,rqd} = \frac{0.87 f_y}{f_{bu}}\frac{\phi}{4} = \frac{0.87(500)}{2.96}\frac{10}{4}$$

$$= 368 \text{ mm}$$

Note that the anchorage length calculated as per HKBD 2013 already penetrates an RC wall thickness of 250 mm.

General method

$$l_{b,rqd} = \frac{f_{sd}}{f_{bu}}\frac{\phi}{4} = \frac{86}{2.96}(\frac{10}{4}) = 73 \text{ mm}$$

Detailed method

$$l_{b,rqd} = \frac{f_{sd}}{f_{bu}}\frac{\phi}{4} = \frac{86}{3.44}(\frac{10}{4}) = 63 \text{ mm}$$

(If it is designed to HKBD 2013)

Using Eq. (4.1), $f_y = 500$ MPa

$$l_{b,rqd} = \frac{0.87 f_y}{f_{bu}}\frac{\phi}{4} = \frac{0.87(500)}{2.96}\frac{10}{4}$$

$$= 368 \text{ mm}$$

Note that the anchorage length calculated as per HKBD 2013 already penetrates an RC wall thickness of 250 mm.

(6) Minimum anchorage length – Proposal 3

Using Eq. (4.38a),

$l_{b,min}$

$\geq \alpha_{lb}$

$\times \max\{0.3 l_{b,rqd}; \ 10\phi; 100 \text{ mm} \}$

General method

$l_{b,min} \geq 1.5 \times \max\{0.3(180) =$

$54; 10(10) = 100; 100\} \geq 150 \text{ mm}$

Detailed method

$l_{b,min} \geq 1.5 \times \max\{0.3(154) =$

$46; 10(10) = 100; 100\} \geq 150 \text{ mm}$

α_{lb} is equal to 1.5, assuming that no tests have been carried out on post-installed reinforcements in cracked concrete in accordance with EAD 330087-00-0601 (2018).

Using Eq. (4.38a),

$l_{b,min}$

$\geq \alpha_{lb}$

$\times \max\{0.3 l_{b,rqd}; \ 10\phi; 100 \text{ mm} \}$

General method

$l_{b,min} \geq 1.5 \times \max\{0.3(73) =$

$22; 10(10) = 100; 100\} \geq 150 \text{ mm}$

Detailed method

$l_{b,min} \geq 1.5 \times \max\{0.3(63) =$

$19; 10(10) = 100; 100\} \geq 150 \text{ mm}$

α_{lb} is equal to 1.5, assuming that no tests have been carried out on post-installed reinforcements in cracked concrete in accordance with EAD 330087-00-0601 (2018).

(7) Provided anchorage length

General method

l_b = max $\{l_{b,req}, l_{b,min}\}$ = max$\{180, 150\}$ = 180 mm

Detailed method

l_b = max $\{l_{b,req}, l_{b,min}\}$ = max$\{154, 150\}$ = 154 mm

General method

l_b = max $\{l_{b,req}, l_{b,min}\}$ = max$\{73, 150\}$ = 150 mm

Detailed method

l_b = max $\{l_{b,req}, l_{b,min}\}$ = max$\{63, 150\}$ = 150 mm

(8) Minimum edge distance check for splitting failure (Proposal 5)

Using Eq. (4.42),

c_d=min$\{s/2,c_1,c_s\}$=min$\{100/2,50,50\}$= 50 mm

From Table 4.5* (with drilling aid, compressed air drilled, and $\phi = 10$)

50 mm + 0.02 l_v

*Or based on the ETA-approved technical data of the material.

General method

50 + 0.02 (180) = 54 mm

Detailed method

50 + 0.02 (154) = 53 mm

Using Eq. (4.42),

c_d=min$\{s/2,c_1,c_s\}$=min$\{100/2,50,50\}$= 50 mm

From Table 4.5* (with drilling aid, compressed air drilled, and $\phi = 10$)

50 mm + 0.02 l_v

*Or based on the ETA-approved technical data of the material.

General and detailed methods

50 + 0.02 (150) = 53 mm

(9) Provided cover

$c_d = 55$

$c_d / \phi = 55/10 = 5.5 > 3$*

(*assuming $\alpha_2 = 0.85$, subject to calculation refinement with $\alpha_2 = 1.0$)

(10) Summary of design for post-installed reinforcements

Use 5 T10 @ 100 mm, $A_{s,prov} = 393$ mm² with $l_b = 180$ mm (general) or $l_b = 160$ mm (detailed method), and $c_d = 55$ mm.

Since l_b extends beyond the centreline of the support (250/2 = 125 mm), one does NOT need to check for additional moments caused by eccentricity on the support.

Use 5 T10 @ 100 mm, $A_{s,prov} = 393$ mm² with $l_b = 150$ mm (general or detailed method), and $c_d = 55$ mm.

Since l_b extends beyond the centreline of the support (250/2 = 125 mm), one does NOT need to check for additional moments caused by eccentricity on the support.

Example 3: A moment connection of an extension of a new RC slab connected to an existing RC slab with a lap splicing

Figure 4.13 shows a new RC slab that needs to be connected to the support of an existing end span slab with post-installed reinforcements to form a continuous system. No reversal of loading is expected; hence, the negative moment occurs at the top of the new intermediate support. The top reinforcement is to be tension lap-spliced to provide adequate anchorage, whilst the bottom can be free from splices but may experience compression in anchorage length due to bending stress, apart from tension caused by shear stress.

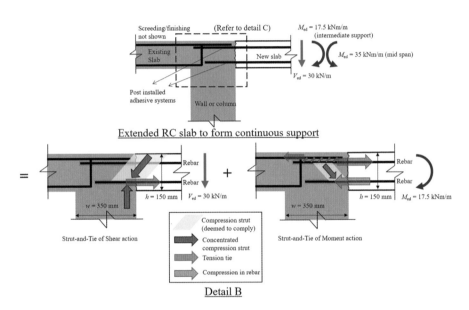

Figure 4.13: Details of end support using post-installed rebar system for continuous supported RC slab, with superposition of shear and moment actions

Structure dimensions, material, and load

Slab: $l_n = 3$ m, $h_{slab} = 150$ mm, $b = 1000$ mm (for per metre run), cover = 30 mm, $d = 120$ mm, and $a_v = d$

Wall or column support: $w = 350$ mm, cover = 50 mm

Concrete strength grade: C40 (cube), $f_{ctk,0.05} \approx 2.1$ MPa

Reinforcement: $f_{yk} = 500$ N/mm^2, $\gamma_s = 1.15$

Structural analysis (ultimate design forces): At mid-span, $M_{Ed} = 35$ kNm/m

At support, $V_{Ed} = 30$ kN/m

Predesigned slab

Bottom reinforcement required: At mid span, $A_{s,rqd,m} = M_{Ed} / (0.9d\ f_{yk}/ \gamma_s) = 745$ mm²/m

Reinforcement provided: At mid span, $\phi10$, $s = 100$ mm; $A_{s,prov,m} = 785$ mm²/m

Deemed-to-comply check for concrete web crushing – Proposal 4:

Method 1: enhanced shear check

Concrete shear stress capacity,

$$v_c = 0.79 \left(\frac{100A_s}{b_v d}\right)^{\frac{1}{3}} \left(\frac{400}{d}\right)^{\frac{1}{4}} \left(\frac{1}{\gamma_m}\right) \left(\frac{f_{cu}}{25}\right)^{\frac{1}{3}} = 0.76 \text{ MPa}$$

(Table 6.3, HKBD 2013)

Assuming a 45-degree strut relative to the bottom longitudinal tensile reinforcement

$2d / a_v = 2$

Concrete enhanced shear stress capacity,

vc enhancement at support $= 0.76(2) = 1.52$ MPa $<$ lesser of ($0.8\sqrt{f_{cu}} = 5.1$ MPa, 7 MPa)

(Cl. 6.1.2.5(g), HKBD 2013)

Applied shear stress,

$V_{Ed} = 30$ kN/m

$v_{Ed} = \frac{V_{Ed}}{b_v d} = 0.20$ MPa < 1.52 MPa, and thus acceptable.

Method 2: strut check using the STM

Assuming a 45-degree strut relative to the bottom longitudinal tensile reinforcement

$F_o = |V_{Ed}| / \sin 45° = 43$ kN/m

Assuming the outer strut width reaches the intersection of the top of the slab with the face of the supporting wall via a 45-degree angle, hence the strut width is $w_{strut} = h/\sqrt{2} = 150/\sqrt{2} = 106$ mm.

$f_{strut} = \frac{F_{strut}}{b w_{strut}} = 0.40$ MPa $< 0.60\ f_c' / 1.5 = 0.50\ f_{cu} / 1.5 = 13.3$ MPa, and thus

acceptable.

(Su, R.K.L and Looi, D.T.W. (2016). "Revisiting the Unreinforced Strut Efficiency Factor", ACI Structural Journal, 113(2), pp. 301–312.)

Check for the anchorage length at the top and bottom post-installed reinforcements at the support, as shown in Figure 4.14.

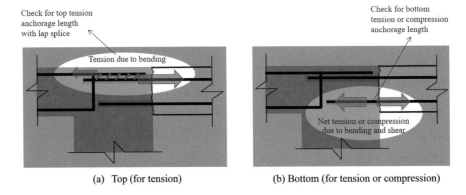

(a) Top (for tension)   (b) Bottom (for tension or compression)

Figure 4.14: Design check for anchorage length of post-installed reinforcement for continuous supported RC slab at intermediate support

Top support	Bottom support
(1) Bond stress – Proposal 1	

The top support is always in tension, whilst the bottom support can be in tension or compression. Bond stress is calculated for both tension and compression.

General method with Eq. (4.35a)

$f_{bu} = \beta \sqrt{f_{cu}} = 0.5\sqrt{40} = 3.16$ MPa (tension)

$f_{bu} = \beta \sqrt{f_{cu}} = 0.63\sqrt{40} = 3.98$ MPa (compression)

Note that the partial safety factor for bond stress = 1.4 is embedded.

Detailed method with Eqs. (4.6), (4.9) and (4.35b)

Although the cover can be equal to 3 ϕ for $\alpha_2 = 0.70$ to maximise the bond strength, it is, however, conservatively equal to 2 ϕ, which implies $\alpha_2 = 0.85$ to account for drilling deviations, as per the recommendation in EAD 330087-00-0601 (2018); reinforcement position is good during concreting and reinforcement diameter is less than 32 mm.

$$f_{bu} = \frac{f_{bd}}{\alpha_2} = 2.25\ \eta_1\eta_2 f_{ctd}/\alpha_2 = 2.25\ \eta_1\eta_2 f_{ctk,0.05}/\gamma_m/\alpha_2$$

$$= 2.25(1)(1)(2.1)/1.5/0.85 = 3.71 \text{ MPa}$$

Compression is not explicitly considered in bond strength, as per EN 1992-1-1 (2004), but directly factored in determining the minimum and required anchorage lengths.

(2) Tension tie force to be anchored – Proposal 2

Shear (V_{Ed}) will not transfer tension to the top bar. The negative moment (M_{Ed}) is equal to 17.5 kNm/m and assumes the effective lever arm (z) is about 90% of the effective depth to form the tension tie force.

Using Eq. (4.36):

$$f_{sd} = F_{Ed}/A_s = [|V_{Ed}| \pm M_{Ed}/z] / A_s$$

Hence, $F_{Ed,tens} = 0 + 17.5/(0.9 \times 0.12) = 162$ kN/m

Post-installed steel area required:

$$A_{s,rqd} = F_{Ed} / (f_{yk} / \gamma_s) = 373 \text{ mm}^2/\text{m}$$

Shear (V_{Ed}) will cause direct tension via a strut of 45-degrees to the bottom bar. The negative moment (M_{Ed}) is equal to 17.5 kNm/m and assumes the effective lever arm (z) is about 90% of the effective depth to form compression force to counter react with the tie force.

Using Eq. (4.36):

$$f_{sd} = F_{Ed}/A_s = [|V_{Ed}| \pm M_{Ed}/z] / A_s$$

Hence, $F_{Ed,tens} = 30 - 17.5/(0.9 \times 0.12) = -132$ kN/m (in compression)

Post-installed steel area required:

$$A_{s,rqd} = F_{Ed} / (f_{yk} / \gamma_s) = 304 \text{ mm}^2/\text{m}$$

(3) Minimum post-installed reinforcement to be anchored at support – Proposal 2

Design as a full moment support (i.e., 50% of the maximum bending in the span) for a continuous slab:

$A_{s,simplified\ rules} = 0.50\ M_{Ed,\ mid\text{-}span} / (0.9d f_{yk}/ \gamma_s) = 373 \text{ mm}^2/\text{m}$

(Cl. 9.3.1.3, HKBD 2013)

At the intermediate support of a continuous slab, 40% of the calculated mid-span bottom reinforcement should be anchored:

$A_{s,simplified\ rules} = 0.40\ A_{s,mid\text{-}span} = 314 \text{ mm}^2/\text{m}$

(Cl. 9.3.1.4, HKBD 2013)

Tension reinforcement, rectangular section subjected to flexure, and web in tension: For a rectangular slab with $f_{y,k} = 500$ MPa,

$A_{s,min} = 0.13\% A_c = 0.13/100\ (1000 \times 150) = 195 \text{ mm}^2/\text{m}$ (Table 9.1, HKBD 2013)

(4) Decision for post-installed reinforcements to be anchored

$A_{s,req} = \max \{373, 373, 195\} = 373 \text{ mm}^2$

Reinforcement provided: 5 nos. $\phi 10$, $s = 200$ mm; $A_{s,prov,m} = 393 \text{ mm}^2/\text{m}$

$A_{s,req} = \max \{304, 314, 195\} = 314 \text{ mm}^2$

Reinforcement provided: 5 nos. $\phi 10$, $s = 200$ mm; $A_{s,prov,m} = 393 \text{ mm}^2/\text{m}$

(5) Anchorage length required (with consideration of lap length)

(Design for Proposal 2)

General method

Using Eq. (4.7), $f_{sd} = 16200 / (5 \pi 10^2/4) = 413$ MPa

$$l_{b,rqd} = \frac{f_{sd}}{f_{bu}} \frac{\phi}{4} = \frac{413}{3.16} (\frac{10}{4}) = 326 \text{ mm}$$

(Design for Proposal 2)

General method

Using Eq. (4.7), $f_{sd} = -13200 / (5 \pi 10^2/4) = -336$ MPa

$$l_{b,rqd} = \frac{f_{sd}}{f_{bu}} \frac{\phi}{4} = \frac{-336}{3.16} (\frac{10}{4}) = 265 \text{ mm}$$

Since a tension lap length is required, an increment factor of 1.4 should be applied:

$l_{b,lap} = 1.4 \times l_{b,rqd} = 1.4 \times 326 = 457$ mm

Detailed method

Using Eq. (4.7), $f_{sd} = 413$ MPa

$l_{b,rqd} = \dfrac{f_{sd}}{f_{bu}} \dfrac{\phi}{4} = \dfrac{413}{3.71} (\dfrac{10}{4}) = 278$ mm

$l_{b,lap} = 1.4 \times l_{b,rqd} = 1.4 \times 278 = 390$ mm

(If it is designed to HKBD 2013)

Using Eq. (4.1), $f_y = 500$ MPa

$l_{b,rqd} = \dfrac{0.87 f_y}{f_{bu}} \dfrac{\phi}{4} = \dfrac{0.87(500)}{3.16} \dfrac{10}{4}$

$= 344$ mm

$l_{b,lap} = 1.4 \times l_{b,rqd} = 1.4 \times 344 = 481$ mm

Note that the anchorage length calculated as per HKBD 2013 requires a longer hole to be drilled.

(compression anchorage)

Detailed method

Using Eq. (4.7), $f_{sd} = -336$ MPa

$l_{b,rqd} = \dfrac{f_{sd}}{f_{bu}} \dfrac{\phi}{4} = \dfrac{-336}{3.98} (\dfrac{10}{4}) = 211$ mm

(note that the compression bond stress is not explicitly considered, therefore the general method in HKBD 2013 is conservatively used here)

(If it is designed to HKBD 2013)

From Eq. (4.1), $f_y = 500$ MPa

$l_{b,rqd} = \dfrac{0.87 f_y}{f_{bu}} \dfrac{\phi}{4} = \dfrac{0.87(500)}{3.98} \dfrac{10}{4}$

$= 273$ mm

Note that the anchorage length calculated as per HKBD 2013 requires a longer hole to be drilled.

(6) Minimum anchorage length – Proposal 3

Using Eq. (4.15) for lapping,

$l_{o,min} \geq \alpha_{lb} \times$

$\max\{0.3\alpha_6 l_{b,rqd}; 15\phi; 200 \text{ mm}\}$

General method

$l_{b,min} \geq 1.5 \times \max\{0.3(1.4)(457) =$

$192; 15(10) = 150; 200\} \geq 300$ mm

Detailed method

$l_{b,min} \geq 1.5 \times \max\{0.3(1.4)(390)$

$= 164; 15(10)$

$= 150; 200\}$

≥ 300 mm

α_{lb} is equal to 1.5, assuming that no tests are carried out for post-installed reinforcements in cracked concrete in accordance with EAD 330087-00-0601 (2018).

Using Eq. (4.38b) for compression,

$l_{b,min}$

$\geq \alpha_{lb} \times \max\{0.6 l_{b,rqd}; 10\phi; 100 \text{ mm}\}$

General method

$l_{b,min} \geq 1.5 \times \max\{0.6(265) =$

$159; 10(10) = 100; 100\} \geq 239$ mm

Detailed method

$l_{b,min} \geq 1.5 \times \max\{0.6(211)$

$= 127; 10(10)$

$= 100; 100\}$

≥ 191 mm

α_{lb} is equal to 1.5, assuming that no tests are carried out for post-installed reinforcements in cracked concrete in accordance with EAD 330087-00-0601 (2018).

α_6 is equal to 1.4, considering only the top reinforcements are lapped and not the bottom reinforcement at the section being considered, hence ρ_1 is taken as 50% in Eq. (4.14).

(7) Provided anchorage length

General method

$l_b = \max \{l_{b,req}, l_{o,min}\} = \max\{457, 300\} =$ 457 mm

Detailed method

$l_b = \max \{l_{b,req}, l_{o,min}\} = \max\{390, 300\} =$ 390 mm

General method

$l_b = \max \{l_{b,req}, l_{b,min}\} = \max\{265, 239\} =$ 265 mm

Detailed method

$l_b = \max \{l_{b,req}, l_{b,min}\} = \max\{211, 191\} =$ 211 mm

(8) Minimum edge distance check for splitting failure (Proposal 5)

Using Eq. (4.42),

$c_d = \min\{s/2, c_1, c_s\} = \min\{100/2, 30, 30\} =$ 30 mm

From Table 4.5 (with drilling aid, compressed air drilled, and $\phi = 10$)

50 mm + 0.02 l_v

General method

50 + 0.02 (457) = 60 mm

Detailed method

50 + 0.02 (390) = 58 mm

Using Eq. (4.42),

$c_d = \min\{s/2, c_1, c_s\} = \min\{100/2, 30, 30\} =$ 30 mm

From Table 4.5 (with drilling aid, compressed air drilled, and $\phi = 10$)

50 mm + 0.02 l_v

General method

50 + 0.02 (265) = 56 mm

Detailed method

50 + 0.02 (211) = 54 mm

(9) Provided cover

$c_d = 60$
$c_d / \phi = 60/10 = 6 > 3*$
(*assuming $\alpha_2 = 0.85$, subject to calculation refinement with $\alpha_2 = 1.0$)

(10) Summary of design for post-installed reinforcements

Use 5 T10 @ 100 mm, $A_{s,prov} = 393$ mm² with $l_b = 460$ mm (general method) or $l_b = 390$ mm (detailed method), and $c_d = 60$ mm.

Use 5 T10 @ 100 mm, $A_{s,prov} = 393$ mm² with $l_b = 265$ mm (general method) or $l_b = 215$ mm (detailed method), and $c_d = 60$ mm.

Example 4: A moment connection of an RC slab on a wall (based on the state-of-the-art design method that uses the STM)

In A&A work, additional slabs may need to be installed on existing walls. Hence, post-installed reinforcements might be used for moment connections. Figure 4.15 shows the moment connection designed for a continuous RC slab with a width of 500 mm and a thickness of 200 mm.

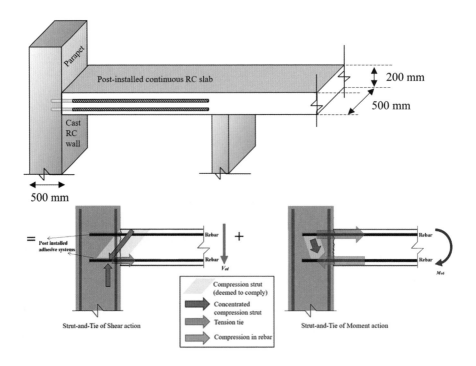

Figure 4.15: Construction of new continuous slab by post-installed rebar system to existing wall structures

Structure dimensions, material, and load

Slab: $h_{slab} = 200$ mm, $b = 500$ mm, clear cover = 20 mm, and $d = 165$ mm,

Wall: thickness = 500 mm, clear cover = 35 mm, reinforcement = 4 ϕ32 at NF, 4 ϕ32 at FF

Concrete strength grade: $f_{cu} = 35$ MPa, $f_{ctk} = 1.95$ MPa, $\gamma_c = 1.5$

Reinforcement: $f_{yk} = 500$ MPa, $\gamma_s = 1.15$

Design point load combination: At ULS, $S_d = (1.40\ G_k + 1.60\ Q_k) = 60$ kN

Design forces: At support, $V_{Ed} = 60$ kN

At support, $M_{Ed} = 45$ kNm

Predesigned slab

Top reinforcement required: $K = M_{Ed}/(bd^2 f_{cu}) = 0.0945$

$$z = d\,[0.5 + (0.25 - K/0.9)^{0.5}] = 0.88\,d = 145 \text{ mm}$$

$$A_{s,rqd} = M_{Ed} / (z\,f_{yk}/\gamma_s) = 712 \text{ mm}^2$$

Reinforcement provided: 3 nos. $\phi20@165$ mm; $A_{s,prov} = 942$ mm^2

Deemed-to-comply check for concrete web crushing – Proposal 4:

Method 1: enhanced shear check

Concrete shear stress capacity,

$$v_c = 0.79 \left(\frac{100 A_s}{b_v d}\right)^{\frac{1}{3}} \left(\frac{400}{d}\right)^{\frac{1}{4}} \left(\frac{1}{\gamma_m}\right) \left(\frac{f_{cu}}{25}\right)^{\frac{1}{3}} = 0.92 \text{ MPa} \qquad \text{(Table 6.3, HKBD 2013)}$$

$$< \text{lesser of } (0.8\sqrt{f_{cu}} = 4.7 \text{ MPa}, 7 \text{ MPa})$$

$$\text{(Cl. 6.1.2.5(g), HKBD 2013)}$$

Applied shear stress,

$V_{Ed} = 60$ kN

$v_{Ed} = \dfrac{V_{Ed}}{b_v d} = 0.73$ MPa < 0.92 MPa, and thus acceptable.

Check for the anchorage length at the top and bottom post-installed reinforcements at the support.

Top support	Bottom support
(1) Bond stress – Proposal 1	

General method using Eq. (4.35a)

$f_{bu} = \beta\sqrt{f_{cu}} = 0.5\sqrt{35} = 2.96$ MPa

Note that the partial safety factor for bond stress = 1.4 is included.

Detailed method using Eq. (4.35b)

In this case, drilling deviations have no significant effect on wall-slab connections. Furthermore, reinforcement position is good during concreting and reinforcement diameter is less than 32 mm (EAD 330087-00-0601 (2018)).

As an example, the characteristic resistance for bonded anchor $\phi20$ reinforcements under tension in concrete for a particular injection system with product performance approved by ETA is $f_{bd,p} = 9.3$ N/mm^2

$c_d = \min\left\{\frac{s}{2}, c_1, c_s\right\} = \min\left\{\frac{165}{2}, 75, \gg 75\right\} = 75$ mm

For reinforcements with a diameter of 20 mm

$$\alpha_2 = 1 - \frac{0.15(75-20)}{20} = 0.59 \le 0.7,$$

use 0.7 (c_d / ϕ is taken as 3.0) to preclude splitting failure (Proposal 5)

using Eq. (4.6)

$$f_{bu} = \frac{f_{bd}}{\alpha_2} = (2.25\,\eta_1\eta_2 f_{ctd})/\alpha_2 = (2.25\,\eta_1\eta_2 f_{ctk}/\gamma_m)/\alpha_2$$

$$= \frac{2.25(1)(1)(1.95)}{1.5}\frac{1}{0.7} = 4.18 \text{ MPa}$$

(2) Tension tie force to be anchored – Proposal 2

Shear (V_{Ed}) will not transfer tension to the top bar. The negative moment is supported by 88% of the effective depth to form the tensile force.

$$f_{sd} = F_{Ed}/A_s = [|V_{Ed}| \pm M_{Ed}/z] / A_s$$

$F_{Ed,tens} = 0 + 45/(0.88 \times 0.165) = 310 \text{ kN}$

Post-installed reinforcement required:

$A_{s,rqd} = F_{Ed} / (f_{yk} / \gamma_s) = 712 \text{ mm}^2$

The bottom support is in compression and compression steel is not required. Hence, 50% of the required tension steel is adopted for the post-installed reinforcement.

$A_{s,rqd} = 712/2 = 366 \text{ mm}^2$

(3) Minimum post-installed reinforcements to be anchored at the support – Proposal 2

Tension reinforcement, rectangular section subjected to flexure, and web in tension:
For rectangular slab with $f_{y,k} = 500$ MPa,

$A_{s,min} = 0.13\% A_c = 0.13/100 \ (500 \times 200) = 130 \text{ mm}^2$ (Table 9.1, HKBD 2013)

(4) Decision for post-installed reinforcements to be anchored

$A_{s,req} = \max \{712, 130\} = 712 \text{ mm}^2$

Reinforcement provided:

3 nos. $\phi 20$; $A_{s,prov} = 942 \text{ mm}^2$

$A_{s,req} = \max \{366, 130\} = 366 \text{ mm}^2$

Reinforcement provided:

3 nos. $\phi 16$; $A_{s,prov} = 603 \text{ mm}^2$

(5) Anchorage length required

(Design for Proposal 2)
Using Eq. (4.7)
$f_{sd} = 310000 / (3 \pi 20^2/4) = 329$ MPa

General method

$$l_{b,rqd} = \frac{f_{sd}}{f_{bu}}\frac{\phi}{4} = \frac{329(20)}{2.96(4)} = 555 \text{ mm}$$

(Design for Proposal 2)
Using Eq. (4.7)
$f_{sd} = 0$ MPa

General method

N.A.

Detailed method

$l_{b,rqd} = \dfrac{f_{sd}}{f_{bu}}\dfrac{\phi}{4} = \dfrac{329\,(20)}{4.18\,(4)} = 393$ mm

(If it is designed to HKBD 2013)

Using Eq. (4.43a), $f_y = 500$ MPa

$l_{b,rqd} = \dfrac{0.87 f_y}{f_{bu}}\dfrac{\phi}{4} = \dfrac{0.87(500)}{2.96}\dfrac{20}{4}$

$= 735$ mm

Note that only the anchorage length calculated as per the detailed method is within a wall thickness of 500 mm.

Detailed method

N.A.

(If it is designed to HKBD 2013)

Using Eq. (4.43a), $f_y = 500$ MPa

$l_{b,rqd} = \dfrac{0.87 f_y}{f_{bu}}\dfrac{\phi}{4} = \dfrac{0.87(500)}{2.96}\dfrac{16}{4}$

$= 588$ mm

Note that the anchorage length calculated as per the codified method is larger than the wall thickness of 500 mm.

(6) Minimum anchorage length – Proposal 3

Using Eq. (4.38a),

$l_{b,min} \geq \alpha_{lb} \times$

$\max\{0.3 l_{b,rqd}, 10\phi, 100\text{ mm}\} =$

$1.5 \times \max\{118, 200, 100\text{ mm}\} =$

300 mm

α_{lb} is equal to 1.5, assuming that no tests are carried out for post-installed reinforcements in cracked concrete in accordance with EAD 330087-00-0601 (2018).

Using Eq. (4.38b),

$l_{b,min} \geq \alpha_{lb} \times$

$\max\{0.6 l_{b,rqd}, 10\phi, 100\text{ mm}\}$

$= 1.5 \times \max\{0, 160, 100\} \geq 240$ mm

α_{lb} is equal to 1.5, assuming that no tests are carried out for post-installed reinforcements in cracked concrete in accordance with EAD 330087-00-0601 (2018).

(7) Provided anchorage length

Detailed method

$l_b = \max\{l_{b,req}, l_{b,min}\} =$

$\max\{393, 300\} = 393$ mm

Provide an anchorage length of 400 mm.

Detailed method

$l_b = \max\{l_{b,req}, l_{b,min}\} =$

$\max\{0, 240\} = 240$ mm

Provide an anchorage length of 240 mm.

(8) Minimum edge distance check for splitting failure (Proposal 5)

From Eq. (4.42),

$c_d = \min\{\frac{s}{2}, c_1, c_s\} = \min\{83, 75, >\!>75\}$

$c_d = c_1 = 75$ mm (side cover)

with drilling aid, hammer drilled, and $\phi = 20$

30 mm + 0.02 $l_v \geq 2\phi$

with drilling aid, hammer drilled, and $\phi = 16$

30 mm + 0.02 $l_v \geq 2\phi$

*Or based on the ETA-approved techni-cal data of the material.

$30 + 0.02 (400) = 38$ mm $< 2\phi = 40 <$ 75 mm side cover is acceptable.

*Or based on the ETA-approved techni-cal data of the material.

$16 + 0.02 (240) = 35$ mm $< 2\phi = 33 <$ 75 mm side cover is acceptable.

(9) Provided cover

$c_d = 75$

$c_d / \phi = 75/20 = 3.75 \geq 3*$

(*assuming $\alpha_2 = 0.7$)

(10) Reinforcement check for near face of wall based on the STM (Proposal 6)

Using $l_{b1} = 400$ mm, as determined in (7)

Using Eqs. (4.19a) and (4.22),

$z_0 = l_{b1}/2 = 200$ mm

$l_{bn} = c_s + \phi/2 + z_0 + l_{b1}/2 = 35 + 32/2 + 200 + 400/2 = 451$ mm

$z_{1r} = z = 145$ mm

$\theta = \tan^{-1} (z_0 / z_{1r}) = \tan^{-1} (200/145) = 54° > 30°$ is acceptable.

$z_{wall} = 415$ mm

$F_{s0} = M_{Ed} (1/z_0 - 1/z_{wall}) = 45,000 (1/200 - 1/415) = 117$ kN

$A_{s,rqd} = F_{s0}/(f_{yk} /\gamma_s) = 117 \times 10^3/(500/1.15) = 268$ mm^2

$A_{s,prov} = 4T32$ (3217 mm^2) > 268 mm^2 is acceptable

(11) Reinforcement check for far face of wall based on the STM

Using Eq. (4.23),

$F_{s3} = M_{Ed}/z_{wall} = 45,000/415 = 108$ kN

$A_{s,rqd} = F_{s3}/(f_{yk} / \gamma_s) = 108,000/(500/1.15) = 249$ mm^2

$A_{s,prov} = 4T32$ (3217 mm^2) > 249 mm^2 is acceptable

(12) Compressive strut check based on the STM

Using Eq. (4.24),

$$F_{c0} = \frac{M_{Ed}}{z_0} = \frac{45,000}{200} = 225 \text{ kN}$$

Using Eq. (4.27),

$$F_0 = \frac{F_{c0}}{\cos\theta} = \frac{225}{\cos 54} = 383 \text{ kN}$$

Using Eq. (4.28),

$F_R = \alpha_{cc} f_{ck} /\gamma_c (b \, l_{b1} \cos \theta)$

$= 0.75\eta_{fc} \, f_{ck} /\gamma_c (b \, l_{b1} \cos \theta)$

$= 0.75 \times 1.0 \times (0.8\times35 / 1.5) (500 \times 400 \cos 54)/1000$

$= 1646$ kN ≥ 383 kN, which is acceptable.

(13) Splitting tensile stress in discontinuity zone based on the STM

Using Eq. (4.29),

$$f_{sp} = F_{c0} \cdot z_0 \times \left(1 - \frac{z_0}{z_{wall}}\right) \times \left(1 - \frac{l_{b1}}{2z_{wall}}\right) / \frac{b \cdot z_{wall}^2}{2.42}$$

$$f_{sp} = 225{,}000 \times 200 \times \left(1 - \frac{200}{415}\right) \times \left(1 - \frac{400}{2 \times 415}\right) / \frac{500\,(415)^2}{2.42}$$

$$f_{sp} = 0.34 \text{ MPa}$$

Using Eq. (4.34),

$f_{sp} \le \alpha_{ct} f_{ctk,0.05} / \gamma_c = 1.0 \times 1.95 / 1.5 = 1.29$ MPa, which is acceptable.

(14) Summary for design of post-installed reinforcements

Using 3T20 top and 3T16 bottom post-installed reinforcements with the anchorage depths l_{bn} = 451 mm and 240 mm, respectively, clear cover at the wall edge = 75 mm is acceptable.

Lee et al. (2019) show that confinement can be used to achieve a higher bond strength of the post-installed reinforcements, which is associated with pull-out failure. Taking into consideration the confinement effect, the anchorage length can be further reduced.

Use confinement method (extension of method in EN 1992-1-1 (2004))

f_{bd} = 2.29 MPa   Using Eq. (4.6a)

As c_d (= 75 mm) is greater than 3ϕ, pull-out failure is prevalent. Hence, the confinement method is applicable.

For a specific adhesive product, and based on its own tests, δ is found to be 0.3. The characteristic bond strength associated with pull-out failure $f_{bd,p}$ is 9.3 (as approved by ETA).

$$\alpha_2' = \frac{1}{\frac{1}{0.7} + \delta \cdot \frac{c_d - 3\phi}{\phi}} = \frac{1}{\frac{1}{0.7} + 0.3 \cdot \frac{75 - 3(20)}{20}} = 0.604 \qquad \text{Using Eq. (4.30)}$$

$$f_{bd,sp} = \frac{f_{bd}}{\alpha_2'} = \frac{2.29}{0.6} = 3.79 \text{ MPa}$$

$$f_{bu} = \min\{f_{bd,sp}, f_{bd,p}\} = \min(3.79, 9.3) = 3.79 \text{ MPa}$$

$$l_{bd,sp} \ge \max\left\{\frac{f_{sd}}{f_{bu}} \frac{\phi}{4}, l_{b,min}\right\} = \{\frac{329}{3.79} \frac{20}{4}, 300\} = 434 \text{ mm}$$

Hence, the anchorage depth can be further reduced from l_{bn} = 451 mm to 434 mm.

5
Qualification and Quality Control of Post-installed Reinforcement Connections

The performance of post-installed reinforcement systems is greatly affected by the type of adhesive used (see Chapter 2), drilling and installation methods (Chapter 3), and the designed length and diameter of the holes (Chapter 4). Hence, product qualification procedures should be in place to ensure that the performance of post-installed reinforcement connections is comparable to that of monolithic cast in-situ connections.

As shown in Tables 1.1 and 4.2a, the two main sources of reference for qualifications are the EAD 330087-00-0601 (2018) developed by the EOTA in Europe and AC 308 (2016) by the ACI in the US. This chapter discusses the overall requirements of these two documents in terms of tests and assessment procedures. The differences between the two documents and their commonalities are also identified and elaborated.

5.1   Qualification of system for post-installed reinforcements

5.1.1 Basic principles

In the EAD 330087-00-0601 (2018) and AC 308 (2016), the primary goal of qualification under static conditions is to establish a comparable performance of cast-in reinforcements with respect to the failure modes. Anticipated failure modes (i.e., bond and splitting) have been discussed in Chapter 4 (see Figures 4.5 and 4.8). The comparison in terms of the load-displacement behaviour (i.e., stiffness) is largely based on extensive research work carried out by Spieth (2002).

The basic tension test procedure to derive the average bond strength of a post-installed reinforcement system under various conditions (i.e., dry or wet conditions; different temperatures, directions, and depths; or in corrosive, alkaline, or sulphuric environments) are described in EAD 330087-00-0601 (2018) and AC 308 (2016). Readers are encouraged to refer to these two sources for details.

5.1.2 Applications of EAD 330087-00-0601 (2018) and AC 308 (2016)

Table 5.1 provides a general comparison of the relevant applications of post-installed reinforcing bars (Genesio et al., 2017a). Post-installed reinforcement systems subjected to static conditions are commonly outlined in both the EAD 330087-00-0601 (2018) and AC 308 (2016). The former provides optional testing provisions to assess the exposure of the product to fire whilst the latter provides qualification for seismic conditions. Both have not offered qualification under fatigue conditions. Interestingly, the concrete cover in the EAD 330087-00-0601 (2018) is required to be larger than the cast-in reinforcement design, whereas AC 308 (2016) allows the same cover thickness provided, of which has been validated through testing.

Table 5.1: General comparison of relevant applications of EAD 330087 (2018) and AC 308 (2016)

Condition	EAD 330087-00-0601 (2018)	AC308 (2016)
Under static conditions	YES	YES
Under seismic conditions	Refer to EAD 331522-00-0601 (endorsed draft 2018)	YES
Under fatigue conditions	NO	NO
Exposure to fire	YES	NO
Concrete cover controls	Larger cover than cast-in rebar design in EN 1992-1-1 (2004)	If verified by test, same cover as cast-in rebars in accordance with ACI 318 (2014)

5.1.3 Specific differences between EAD 330087-00-0601 (2018) and AC 308 (2016)

Despite that EAD 330087-00-0601 (2018) and AC 308 (2016) were both largely developed based on the same research findings, they have significant differences. Table 5.2 lists the specific differences between the two documents.

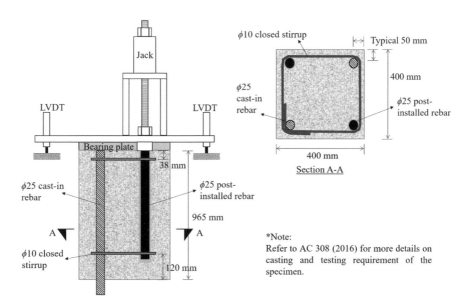

Figure 5.1: Example of splitting test setup, as per AC 308 (2016)

Table 5.2: Specific differences between EAD 330087-00-0601 (2018) and AC 308 (2016)

Test assessment	EAD 330087-00-0601 (2018)	AC 308 (2016)
Adhesive bond strength	Limit in accordance with $$f_{bm}{}^{req} = \frac{\gamma_c f_{bd}}{0.75}\left(\frac{40}{10}\right)^{0.55}$$ for concrete class up to C50/60, where f_{bm} is the required bond resistance of post-installed systems, γ_c is the material partial factor for concrete equal to 1.5, and f_{bd} is the design value of ultimate bond resistance in accordance with EN 1992-1-1 (2004) for good bond conditions. If the product exhibits lower average bond strength but has at least f_{bm} = 7.1 MPa, concession is allowed for qualification, with a bond strength reduction factor.	No concession. All limits must be equal or exceed 7.5 MPa and 11.8 MPa for low and high strength uncracked concrete, respectively.
Splitting failure test	None.	Additional test required for bond/splitting behaviour. This is to avoid 'zipper' failure from excessive adhesive stiffness due to shear lag at the corner or near-edge bars. On the contrary, avoid overly 'soft' adhesives, which lead to relaxation, low stiffness, cracks, and corrosion (see Fig. 5.1 for example of test setup).
Cracked concrete test	Optional. Mainly to avoid using long development lengths. Bond strength is reduced by approximately 25% observed in 0.3 mm longitudinal cracks for cast-in bars (Eibl et al., 1997). Performance of post-installed rebars in cracked concrete is assumed to be 50% of uncracked concrete.	Checking the bond strength and displacement is mandatory. Bond strength is reduced by approximately 50% observed in 0.4 mm cracks (Simons, 2007). No assumption of preliminary reduction is made on the performance of post-installed systems.
Minimum edge distance and concrete cover	Minimum edge distance and concrete cover are required, depending on drilling method, bar diameter, and use of drilling aid (see Table 4.5).	Same concrete cover as that for cast-in rebar design in ACI 318 (2014).
Installation depth	To be tested at a reachable depth (e.g., 20 ϕ, 80 ϕ, etc.), which allows for more flexibility and product differentiation.	Requires successful installation at a depth of 60 ϕ.

5.1.4 Seismic assessments: Cyclic testing as per EAD 331522-00-0601 (endorsed draft 2018) and AC 308 (2016)

The probability of a large earthquake occurring in Hong Kong is low due to its geographical location. Whilst Hong Kong is not located anywhere near a tectonic boundary, the city is however subjected to low-to-moderate intraplate seismic activity. Therefore, there has been effort made to draft seismic design codes (HKBD, 2017). This section will present the assessment methods for post-installed reinforcements subjected to seismic loads in the EAD 331522-00-0601 (endorsed draft 2018) and AC 308 (2016). Engineers are reminded to first qualify the post-installed reinforcement system under static loading as a pre-requisite before proceeding to conducting a seismic assessment.

EAD 331522-00-0601 (endorsed draft 2018)
The EAD 331522-00-0601 (endorsed draft 2018) provides a detailed method for the seismic assessment of post-installed reinforcement systems by taking both small and large covers into consideration. Seismic bond splitting tests are used for small covers to check that the dissipated energy of a post-installed reinforcement is not lower than that of a cast-in bar. More specifically, if a post-installed reinforcement system does not fulfil this criterion, installation can only proceed with a larger minimum concrete cover that is comparable to the bond strength in case there is seismic loading. This provision applies across other standards and should be used jointly with EN 1992-1-1 (2004) and EN 1998-1 (2004), taking into account the reduction factor (k_b) in Section 2.2.2 and amplification factor for minimum anchorage length (α_{lb}) in Section 2.2.3 of EAD 330087-00-0601 (2018).

The test setup to determine the bond strength under seismic loading is shown in Figure 5.2. The drilled hole should be deep enough to prevent the reinforcement from coming into contact with the end of the hole. A compressible material should be inserted at the bottom of the hole prior to injecting the adhesive and inserting the bar. This confinement test is conducted with a friction reducing material placed between the confining plate and the concrete surface. The objective of the test is to check whether there are any increases in the bond-strength degradation of post-installed reinforcements with increasing numbers of cycles versus cast-in reinforcements. The bond strength for seismic applications ($f_{bd,seis}$) is intended to replace the bond strength for static loading (f_{bd}) in EN 1992-1-1 (2004). The required bond stress under cyclic activity is summarised in the EAD 331522-00-0601 (endorsed draft 2018).

AC 308 (2016)
The testing equipment and specimens used to obtain the seismic resistance of post-installed reinforcements under seismic conditions based on AC 308 (2016) are shown in Figure 5.2 (see p. 82). The setup is similar to that based on the EAD 331522-00-0601 (endorsed draft 2018). Note that splitting failure under cyclic loading is required with the EAD 331522-00-0601 (endorsed draft 2018). The required bond stress under cyclic activity is summarised in AC 308 (2016). Readers are encouraged to refer to these two documents for details.

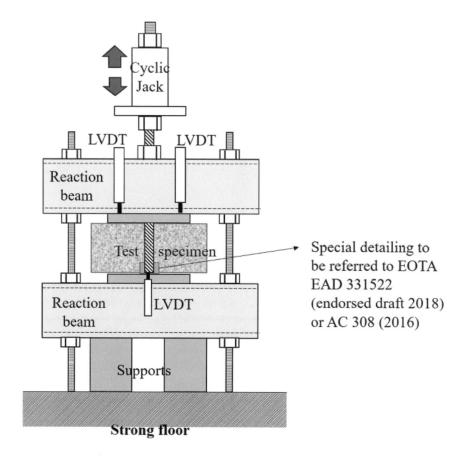

Figure 5.2: Example of confined tension test setup for cyclic testing of reinforcing bar, as per EAD 331522-00-0601 (endorsed draft 2018) and AC 308 (2016)

5.1.5 Fire exposure assessment: EAD 330087-00-0601 (2018)

The fire resistance of post-installed systems is an essential concern in practical applications. When exposed to fire, post-installed systems are expected to demonstrate reduced performance (i.e., bond stress) due to the effects of high temperatures.

The EAD 330087-00-0601 (2018) provides a test for qualifying a post-installed system that has been exposed to fire. Figure 5.3 shows the test setup. The test is to be performed within 14 days after curing the concrete. The minimum heating rate is 5°K/min, with a minimum of 20 test cycles. Readers are encouraged to refer to the EAD 330087-00-0601 (2018) for details.

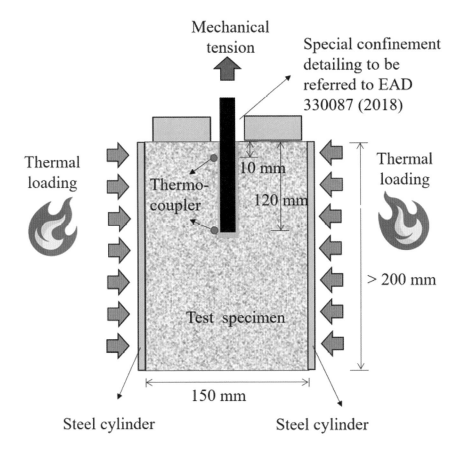

Figure 5.3: Fire resistance of post-installed reinforcing bar connection, as per EAD 330087-00-0601 (2018)

5.1.6 Recommendations for qualifying post-installed reinforcements in Hong Kong

Although all are different documents, the EAD 330087-00-0601 (2018), EAD 331522-00-0601 (endorsed draft 2018) and AC 308 (2016) also have many areas of commonality. Hence, rather than simply adopting one of the guidelines, it is recommended that missing criteria should be incorporated. The recommended points that should be included for the qualification of post-installed reinforcement systems include the following:

    a) All basic tension testing should follow either EAD 330087-00-0601 (2018) or AC 308 (2016).

    b) Splitting tests should follow AC 308 (2016).

c) Cyclic tests should refer to AC 308 (2016) or the more detailed method in the EAD 331522-00-0601 (endorsed draft 2018).

d) Fire resistance testing should follow the EAD 330087-00-0601 (2018).

5.2 Supervision, inspection, and certification

This section provides guidance on the requirements for the supervision, inspection, and certification of post-installed reinforcements on construction sites. In view of the similarities in the installation process of post-installed reinforcements to adhesive anchors, this section offers some amendments of the latter, but the guidelines are still largely based on BS 8539:2012 *Code of Practice for the Selection and Installation of Post-installed Anchors in Concrete and Masonry* (ETAG 001, 2013).

5.2.1 Supervision

Close supervision of the installation of post-installed reinforcements should be undertaken by a supervisor who is a competent member of the site management team. The appointed supervisor should be trained and competent in the installation of post-installed reinforcements.

The supervisor should ensure that the installer is experienced and qualified to use the adhesion system being applied, and should be aware of the consequences of failure to adhere to the correct installation instructions. The supervisor should ensure that the post-installed reinforcements are installed in accordance with the manufacturer's instructions, design criteria, approved construction specifications, and applicable evaluation report.

In smaller projects (or cases where the safety relevance of the post-installed reinforcement system is lower), the role and responsibilities of the supervisor may be undertaken by the installer if given authorisation from the contractor.

In Hong Kong, qualified site supervision of post-installed reinforcements should be carried out by an experienced and competent individual as defined in the following:

(1) The Registered Structural Engineer should assign a quality control supervisor to supervise the works, determine the necessary frequency of inspection by the quality control supervisor (which should not be less than once a week), and devise inspection checklists. The minimum qualifications and experience of the quality control supervisor is to be the same as a Technically Competent Person of Grade T3, as stipulated in the Code of Practice for Site Supervision 2009.

(2) The Registered General Building Contractor should assign a quality control coordinator to provide full-time on-site supervision of the works and devise inspection checklists. The minimum qualifications and experience of the quality control coordinator is to be a Technically Competent Person of Grade T1, as stipulated in the Code of Practice for Site Supervision 2009.

5.2.2 Supervision logbook

The supervisor should ensure that the following issues have been adequately addressed in the supervision logbook:

1. the post-installed reinforcement type that is being used meets the design requirements of the specifier, which include:
   a) type of reinforcement,
   b) reinforcement material specifications, and
   c) reinforcement diameter and length;
2. the reinforcement is positioned in accordance with the design guidelines, which include:
   a) its embedment depth,
   b) its location,
   c) the edge distance, and
   d) its spacing requirements;
3. the base material conditions and the hole dimensions are as specified, which include:
   a) the concrete strength,
   b) reinforcement hole diameter and depth both adhere to the manufacturer's recommendations, and
   c) setting out of the reinforcement on the base material follows design guidelines;
4. the reinforcement is installed with the correct equipment and follows the manufacturer's instructions, which include:
   a) the roughening method, tools, and depth for old concrete surfaces,
   b) hole drilling method and drilling equipment including type of bit and size,
   c) hole cleaning procedures and cleaning tools including brushes, compressed air, steel brush, extensions for deep holes, vacuum cleaner, etc. (see Chapter 3),
   d) the adhesive injection procedures and tools, including extension tube and piston plugs for deep holes,
   e) reinforcement installation procedures and tools including stoppers and end caps for holes drilled overhead, and
   f) cleaned hole and installed reinforcement protection methods;
5. adhesive used should be qualified and the manufacturer's instructions are followed, which include:
   a) adhesive type,
   b) adhesive expiry date,
   c) adhesive storage conditions, and
   d) curing times;
6. the concrete temperature must be confirmed prior to the installation to conform with the requirements of the manufacturer and to establish the curing time for the adhesive; and
7. change in requirements: if amendments are required due to interference with existing reinforcements during drilling, quality of the concrete base material,

etc., a revised installation procedure should be agreed on with the specifier and then communicated to the installer in writing.

The names and qualifications of the Registered Structural Engineer and the Registered General Building Contractor, respectively, should be recorded in the supervision logbook. The date, time, items supervised, and results should also be clearly recorded in the logbook. The logbook should be kept on site for inspection by representatives of the Buildings Department.

5.2.3 Inspection

Immediately following installation and prior to loading, the post-installed reinforcements should be inspected so that any observation of rotation, movement, deformation, cracking, or other types of damage can be recorded and communicated to the specifier.

If, at any stage, the supervisor has any concerns about the suitability of the reinforcement, the placing of further units shall not proceed and the reinforcement in question should be made safe until the concerns have been addressed to the satisfaction of the specifier. A similar inspection should be undertaken at each stage of the subsequent loading.

5.2.4 Certification

The installer and/or supervisor certifies by issuing a certificate that the post-installed reinforcements have been correctly installed in accordance with the specifications and are in a condition to be loaded.

5.3   On-site testing

This section provides guidance on the requirements for the on-site testing of post-installed reinforcements to validate their quality after installation. The performance of post-installed reinforcements can be very different from that anticipated at the design stage, as their performance is heavily influenced by workmanship, as well as environmental and jobsite conditions. Hence, on-site testing of post-installed reinforcements should not be considered as a prequalification process but to verify the workmanship of the installation. Prequalification is also done to confirm the adhesion system (see Section 5.1). The post-installed reinforcements shall be installed by trained installers who are working under qualified supervisors (see Section 5.2).

In view that the installation process of post-installed reinforcements is similar to that of adhesive anchors, some recommendations are proposed based on a few relevant references, namely the BS 8539 (2012), CFA Guidance Note: Procedure for Site Testing Construction Fixing (2012) and the BD 2-3/9018/82/3(s).

5.3.1 Tester

The tester should ensure that the post-installed reinforcements are installed in accordance with the manufacturer's instructions and in the test locations stated by the specifier or other individuals responsible for the supervision of the tests (e.g., the contractor). The tester should record the results comprehensively and in detail, and communicate them to the specifier or responsible individual who is requesting the tests.

Pull-out strength tests should be carried out by a competent tester (independent of the contractor), ideally equipped with knowledge of post-installed reinforcements. The experience of the tester in understanding the workflow and failure modes of the post-installed system will significantly improve the tests and allow for constructive feedback which can be provided to the responsible engineer who is requesting the tests. In Hong Kong, strength tests on a representative number of post-installed reinforcing bars should be carried out as directed by the Registered Structural Engineer.

5.3.2 Testing requirements to validate quality of installation

1. Minimum number

The minimum number of post-installed reinforcements to be tested should be at least 1% or 5 of each type, size, embedment depth, and concrete strength, whichever is more.

2. Test load

Each selected representative post-installed reinforcing bar should be tested for tensile load (by pull-out testing). The test load is recommended to be 1.5 times the working load or otherwise specified by the designer in the drawings. The testing load should be always less than $0.87 f_y$ (reinforcement yield strength with 1.15 as material safety factor).

3. Acceptance criteria

The quality of the post-installed reinforcement is accepted if the test load can be held without any signs of separation, plastic deformation, deleterious effects, movement, or any damage to either the post-installed reinforcement or concrete material. Otherwise, the post-installed reinforcement should be recorded as a failure.

4. Failures during testing

Failures during testing are indicative of unsatisfactory installation processes, which are to be taken seriously. If one post-installed reinforcement fails, the reason for the failure should be investigated, and then the number of tests in that discrete area should be doubled (refer to Section B.3 in BS 8539: 2012 for proof testing of post-installed anchors in concrete). If more than one post-installed reinforcement fails, then all or 100% of the post-installed reinforcements in that group should be tested. Reasons for failure should be communicated to those responsible for the specifications of the installation of the post-installed reinforcements (i.e., designer and installer), as this may require modifications to the installation method.

5. Replacement of failed post-installed reinforcements

Failed post-installed reinforcements will usually need to be replaced. This should be carried out by a competent installer who observes all installation procedures as set out in the manufacturer's instructions and in other related standards. The specifications of the replacement installation should be consulted with the specifier and confirmed by them (i.e., specialist and designer) before the replacement is carried out. It may be possible to install a new fixing in the original hole, depending on the nature of the failure. If the replacement of reinforcing bars is not permitted in the same drilled holes, then the location of new holes should be determined by the specifier.

5.3.3 Test report

All relevant information related to load testing should be accurately recorded in the test report. The individual who is requesting the tests determines whether the test objectives have been met or otherwise, and then the action(s) that needs to be taken.

The following are some of the details that should be recorded in a test report, and additional information may be required depending on the specific circumstances of the test:

1. Administration details: date of test and unique report reference number.
2. Client's name, address, contact name, and position.
3. Name and company of the installer.
4. Name of supervisor.
5. Name of the tester with job title and relevant qualifications.
6. Details of post-installed reinforcements: manufacturer, type, size, spacing, and cover thickness.
7. Test objectives: procedure, required test load, etc.
8. Test location and conditions: details of the location of each test, edge distance, centre spacing, thickness, etc.
9. Base material: type and concrete strength.
10. Installation details: hole diameter, drill bit cutting diameter, depth and direction of the drilled holes, effective embedment depth, hole drilling and cleaning methods, ambient temperature during installation, manufacturer's recommended gelling and curing times, and actual gel gelling and curing times.
11. Testing equipment details: manufacturer, type and load capacity of hydraulic ram, gauge and recorder, date of last calibration, calibrating authority, loading frame (dimension between anchor and closest support), manufacturer and type of torque wrench used.
12. Test results: load applied, movement, and failure mode where applicable.
13. Gauge calibration certificate.
14. Comment statement to the effect that the test complied with this procedure and any exceptions.
15. Comment statement as to whether or not fixing(s) meet the test objective.

5.4   Outlook and future development for quality control of post-installed reinforcements in Hong Kong

Apart from the recommended qualification and quality control items for post-installed reinforcements that are relevant to Hong Kong, some outlook and future development items are shared to conclude this Guide.

In the long run, product certification for adhesives (can be other materials/products related to PIR) may be implemented. To achieve this, the production conformity scheme for adhesives, which stipulates a set of rules and procedures for suppliers of specific adhesive products to comply with, should first be written by the scheme owner. This should be followed by a third-party certification body who conducts a rigorous evaluation on the product manufacturer in accordance with the specifications. A product certificate would then be issued by the certification body for the qualified adhesive. Using an adhesive with a valid product certificate would mean that additional laboratory test reports to determine and justify its quality and performance can be avoided.

References

AC 308 (2016). Post-installed Adhesive Anchors in Concrete Elements. Whittier, California: International Code Council Evaluation Service, Inc. (ICC-ES).

ACI 318 (2014). Building Code Requirements for Structural Concrete (ACI 318-14) and Commentary (ACI 318R-14). Farmington Hills, Michigan: American Concrete Institute.

ACI 355.4 (2011). Qualification of Post-Installed Adhesive Anchors in Concrete (ACI 355.4M-11): An ACI Standard and Commentary. Farmington Hills, Michigan: American Concrete Institute.

ACI 408R (2003). Bond and Development of Straight Reinforcing Bars in Tension (ACI 408R-03). Farmington Hills, Michigan: American Concrete Institute.

BD 2-3/9018/82/3(s) Cementitious of Polymer Based Grouted Bolts/Dowels/Reinforcing Bar Works.

Bilal S. Hamad, Rania AI Hammoud, and Jakob Kunz (2006). Evaluation of Bond Strength of Bonded-in or Post-installed Reinforcement. ACI Structural Journal, 103, 2, 207–218.

BS 1881-124 (2015). Testing Concrete: Methods for Analysis of Hardened Concrete. British Standards Institution (BSI), London, UK.

BS 8110-1 (1997). Structural Use of Concrete, Part 1: Code of Practice for Design and Construction. British Standards Institution (BSI), London, UK.

BS 8539 (2012). Code of Practice for the Selection and Installation of Post-installed Anchors in Concrete and Masonry. British Standards Institution (BSI), London, UK.

BS 6744 (2016). Stainless Steel Bars: Reinforcement of Concrete. Requirements and Test Methods. British Standards Institution (BSI), London, UK.

CEN/TS 1992-4-5 (2009). Design of Fastenings for Use in Concrete—Part 4-1-4-5. Technical Specification, European Committee for Standardisation (CEN), Brussels.

CFA Guidance Note (2012). GN Procedure for Site Testing Construction Fixings. Construction Fixings Association. https://www.the-cfa.co.uk/publications-and-downloads/guidance-notes/

Charney, F. A., Pal, K., and Silva, J. (2013). 'Recommended Procedures for Development and Splicing of Pot-installed Bonded Reinforcing Bars in Concrete Structures'. ACI Structural Journal, 110 (3), 437–446.

Cho, S. S. H., and Chan, S. L. (n.d.) Guide on Design of Post-installed Anchor Bolt Systems in Hong Kong. The Hong Kong Institute of Steel Construction. http://www.hkisc.org/announcement/Anchor_bolt_design_handbook.pdf

CS2:2012. Construction Standard CS2: Steel Reinforcing Bars for the Reinforcement of Concrete. Civil Engineering and Development Department (CEDD), Hong Kong.

DIN 1045-1 (2008). Plain, Reinforced and Prestressed Concrete Structures—Part 1: Design and Construction. German Institute for Standardisation (Deutsches Institut für Normung-DIN), Germany.

Eibl, J., Idda, K., and Lucero-Cimas, H. N. (1997). 'Verbundverhalten bei Querzug (Bond behaviour under shear load)'. Forschungsbericht, Institut fur Massivbau and Baustofftecnologue (MB), Universität Karlsruhe (in German).

El-Reedy, M. A. (2008). Steel-RC Structures: Assessment and Repair of Corrosion. CRC Press, Taylor & Francis Group, Boca Raton, FL, US.

EAD 330087-00-0601 (2018). Systems for Post-installed Rebar Connections with Mortar, European Organisation for Technical Assessment (EOTA).

EAD 330232-00-0601 (2016). Mechanical Fasteners for Use in Concrete. European Organisation for Technical Assessment (EOTA).

EAD 330499-00-0601 (2017). Bonded Fasteners for Use in Concrete. European Organisation for Technical Assessment (EOTA).

EAD 331522-00-0601 (Endorsed Draft 2018). Post-installed Rebar with Mortar under Seismic Action. European Organisation for Technical Assessment (EOTA).

EN 1504-6 (2006). Products and Systems for the Protection and Repair of Concrete Structures: Definitions, Requirements, Quality Control and Evaluation of Conformity. Anchoring of Reinforcing Steel Bar. European Committee for Standardization (CEN).

EN 1992-1-1 (2004). Eurocode 2: Design of Concrete Structures—Part 1-1: General Rules, and Rules for Buildings. European Committee for Standardization (CEN).

EN 1992-4 (2018). Eurocode 2: Design of Concrete Structures—Part 4: Design of Fastenings for Use in Concrete. European Committee for Standardization (CEN).

EN 1998-1 (2004). Eurocode 8: Design of Structures for Earthquake Resistance—Part 1: General Rules, Seismic Actions and Rules for Buildings. European Committee for Standardization (CEN).

EN 14629 (2007). Products and Systems for the Protection and Repair of Concrete Structures. Test Methods. Determination of Chloride Content in Hardened Concrete. European Committee for Standardization (CEN).

ETAG 001 (2013). Guideline for European Technical Approval of Metal Anchors for Use in Concrete, Part Five: Bonded Anchors. European Organisation for Technical Assessment (EOTA).

European Technical Assessment ETA-14/0167 of 28/05/2014, Post-installed Rebar Connections of the Sizes 8 to 16 mm with CV.VSF PRO Injection Mortar.

European Technical Assessment ETA-14/0001 of 12/02/2014, Post-installed Rebar Connections Diameters 8 to 25 mm with Hilti HIT-HY 100 Injection Mortar.

fédération internationale du béton (2011a). fib Bulletin 58. Design of Anchorages in Concrete. DCC Document Competence Center Siegmar Kästl e.K., Germany.

fédération internationale du béton (2011b). fib Bulletin 61. Design Examples for Strut-and-Tie Models. Technical Report by Working Party 1.1-3 in fib Task Group 1.1, Design Applications. DCC Document Competence Center Siegmar Kästl e.K., Germany.

fédération internationale du béton (2013). fib Model Code for Concrete Structures 2010. Wilheim Ernst & Sohn. Berlin, Germany.

Gamache, C. (2017). 'Adhesive Anchor Systems, the Effects of Base Material Temperature during Installation and In-service Use', Structure Magazine, National Council of Structural Engineers Associations (NCSEA).

Genesio, G., Piccinin R., and Silva J. (2017a). Qualification of a System for Post-installed Reinforcing Bars under the Rules Established by EOTA and ICC-ES, 3rd International

Symposium on Connections between Steel and Concrete Stuttgart, Germany, 27–29 September.

Genesio, G., Nerbano, S., Piccinin, R. (2017b) Design of Moment Resisting RC Connections Using Post-installed Reinforcing Bars. 3rd International Symposium on Connections between Steel and Concrete Stuttgart, Germany, 27–29 September.

Hilti (2016). Anchor Fastening Technology Manual, Hilti Hong Kong.

HKBD 2009 (2010). Code of Practice for Site Supervision 2009. Buildings Department (BD), Hong Kong.

HKBD 2011 (2011). Code of Practice for Fire Safety in Buildings 2011. Buildings Department (BD), Hong Kong.

HKBD 2013 (2013). Code of Practice for Structural Use of Concrete 2013. Buildings Department (BD), Hong Kong.

HKBD (2017) Draft Seismic Design Code of Practice, Consultancy Study on Seismic Resistant Design Standards for Buildings in Hong Kong (Consultancy Agreement No. 9OC101). Buildings Department, Hong Kong.

Hoermann-Gast, A., and Olsen, J. 'The Effects of Temperature on Post-installed Adhesive Anchors'. Publications of Concrete and Masonry Anchor Manufacturers Association (CAMA), http://www.concreteanchors.org/publications/ac308_temperature.pdf

IAN 104/15 (2015). Anchorage of Post-installed Fasteners and Reinforcing Bars in Concrete. Interim Advice Note 104/15, Highways England.

Kupfer, H., Münger, F., Kunz, J., and Jähring, A. (2003). 'Nachträglich Verankerte Gerade Bewehrungsstäbe bei Rahmenknotenten' (Post-installed Straight Reinforcing Bars in Frame Joints), *Bauingenieur*, Band 78, 15 pp. (in German).

Lee, A. Y. F., Su, R. K. L., and Chan, R. W. K. (2019). 'Structural Behaviour of Post-installed Reinforcement in Wall-slab Moment Connections'. Engineering Structures, 195, 536–550.

Mahrenholtz C., Eligehausen, R., and Reinhardt H. (2015). Design of Post-installed Reinforcing Bars as End Anchorage or as Bonded Anchor, Engineering Structures, 100, 645–655.

Mailvaganam, N. P., Pye, G. B., and Arnott, M. R. (1998). 'Surface Preparation of the Concrete Substrate'. Construction Technical Update No. 24, Institute for Research in Construction, National Research Council of Canada (NRC CNRC), ISSN 1206–1220.

Michigan Department of Transportation, Field Manual for Concrete Anchoring. First edition. February 2015.

Morgan, R. T. (2015). 'Post-installed Reinforcement'. Structure Magazine, 14–16.

Narayanan, R. S., and Goodchild C. H. (2006). The Concise Eurocode 2. The Concrete Centre, Surrey.

Randl, N., and Kunz, J (2012). Concrete Splitting for Rebars Post-installed with High Bond Adhesives. Proceedings of 4th Int. Symposium Bond in Concrete, Brescia, Italy, 17–20 June.

Simons, I. (2007). Verbundverhalten von eingemörtelten Bewehrungsstäben unter zyklischer Beanspruchung [Bond behaviour of post-installed reinforcing bars subjected to cyclic loading], PhD thesis, University of Stuttgart, 2007 (in German).

Spieth, H. (2002). Tragverhalten und Bemessung von Eingemörtelten Bewehrungsstäben [Structural behaviour and design of bonded reinforcing bars], doctoral thesis, University of Stuttgart (in German).

Su, R. K. L., and Looi, D. T. W. (2016). Revisiting the Unreinforced Strut Efficiency Factor. ACI Structural Journal, 113(2), 301–312.

The Concrete Centre (2015). Concrete Design Guide No. 5: How to Calculate Anchorage and Lap Lengths to Eurocode 2. The Institution of Structural Engineers. 46–53.

TR 51 (1998). Guidance on the Use of Stainless-Steel Reinforcement. Concrete Society Technical Report 51. Surrey, UK.

TR 020 (2004). Evaluation of Anchorages in Concrete Concerning Resistance to Fire. Technical Report 020, European Organisation for Technical Assessment (EOTA).

TR 023 (2006). Assessment of Post-installed Rebar Connections. Technical Report 023, European Organisation for Technical Assessment (EOTA).

TR 045 (2013). Design of Metal Anchors for Use in Concrete under Seismic Actions. Technical Report 045, European Organisation for Technical Assessment (EOTA).

About the Authors

Ray K. L. Su is an associate professor at the University of Hong Kong and a fellow of the Hong Kong Institution of Engineers and the Institution of Structural Engineers. He received his PhD in computational fracture mechanics from the University of Hong Kong in 1996 and has published over 140 SCI journal papers in concrete fracturing, durability of reinforced concrete structures, and seismic resistance design of concrete buildings. Dr Su has delivered keynote speeches at many international conferences, including the International Conference on Innovative Materials, Structures and Technologies 2019, the 26th Assembly Advanced Materials Congress 2019, and the Australian Earthquake Engineering Society Conference 2008. He is an academic editor for the *Journal of Applied Mathematics* and editorial panel member for *Sustainability*, *ICE Proceedings: Structures and Buildings*, *Structural Engineering and Mechanics*, and *Advances in Concrete Construction*.

Daniel T. W. Looi is currently a lecturer and course coordinator for the civil engineering programme at Swinburne University of Technology, Sarawak campus, Malaysia. He is a chartered professional engineer (structural) of Engineers Australia and a member of the Earthquake Committee of the Institution of Engineers Malaysia. He obtained his bachelor's degree in civil engineering from the University of Malaya and his PhD in structural engineering from the University of Hong Kong (HKU). He was a postdoctoral fellow in the Department of Civil Engineering of HKU. Dr Looi has published research works in seismic engineering, concrete mechanics, modular buildings, and fastening technologies, which include post-installed reinforcements. He was the recipient of the HKIE Outstanding Paper Award for Young Researcher/Engineer in 2015.

Yanlong Zhang is currently a postgraduate candidate under the supervision of Associate Professor Ir Dr Ray K. L. Su in the Department of Civil Engineering at the University of Hong Kong. He obtained his MSc (Hons) and BSc (Eng) (Hons) degrees from Shijiazhuang Tiedao University, Shijiazhuang, China. His research interests include durability of RC structures, post-installed rebars, bond strength between rebars and concrete, and particle packing and interaction in a concrete mix.